THE MATHEMATICS
LOVER'S COMPANION

EDWARD SCHEINERMAN

THE MATHEMATICS LOVER'S COMPANION

Masterpieces for Everyone

YALE UNIVERSITY PRESS
NEW HAVEN AND LONDON

Copyright © 2017 by Edward Scheinerman.
All rights reserved.

This book may not be reproduced, in whole or in part, including illustrations, in any form (beyond that copying permitted by Sections 107 and 108 of the U.S. Copyright Law and except by reviewers for the public press), without written permission from the publishers.

Yale University Press books may be purchased in quantity for educational, business, or promotional use. For information, please email sales.press@yale.edu (U.S. office) or sales@yaleup.co.uk (U.K. office).

Printed in the United States of America.

Library of Congress Control Number: 2016946694
ISBN 978-0-300-22300-2 (hardcover: alk. paper)

A catalogue record for this book is available from the British Library.

This paper meets the requirements of ANSI/NISO Z39.48-1992 (Permanence of Paper).

10 9 8 7 6 5 4 3 2

To Rachel and Mordechai

Contents

Preface xv
   Joy xv
   Overview xvii
   How to read a mathematics book xviii
   About the cover xviii
   Acknowledgments xix

Prelude: Theorem and Proof 1
   The last words 4

PART I: NUMBER 5

Prime Numbers 7
   The integers 7
   Factoring 9
   How many? 10
   A constructive approach 13
   A different proof 14
   Two infamous problems 16
   Application to cryptography 17

2 Binary   19

   When in Rome   19

   Unary   20

   A middle ground   21

   Calculation   22

   Extensions   25

3 0.999999999999...   27

   The meaning of decimals   27

   Decimal numbers with infinitely many digits   28

   Let's be reckless   30

4 $\sqrt{2}$ 33

   Rational numbers   33

   The diagonal of a square   34

   Beyond the rational   35

   Constructible numbers   38

   Playing in tune   40

5 i 44

   Another square-root conundrum   44

   Imaginary numbers   46

   Complex numbers   47

   The Fundamental Theorem of Algebra   50

6   π   52

   What is π?   52

   Transcendence*   55

   Relatively prime   56

7   e   60

   Leonhard Euler   60

   An "interesting" number   61

   The deranged hat check clerk   64

   The average gap between primes   66

   A miraculous equation   69

8   ∞   71

   Sets   71

   Infinite sets of unequal size   75

   Transfinite numbers   80

   Weirdness in setland   81

Fibonacci Numbers   86

   Squares and dominoes   86

   The Fibonacci numbers   88

   Sums of Fibonacci numbers   89

   Proof by induction   91

   Combinatorial proof   93

   Ratios of Fibonacci numbers and the golden mean   97

10 **Factorial!** 102
  - Books on the shelf 102
  - Is there a formula? 104
  - A puzzle 107
  - What is 0!? 107

11 **Benford's Law** 109
  - Wild measurements 110
  - Multiplication tables 112
  - Catching crooks cooking books 115
  - Refining the problem with scientific notation 115
  - Yards or feet? 117
  - What's logs got to do with it? 120
  - Collecting the loose ends 123

12 **Algorithm** 124
  - Sorting 125
  - Greatest common divisor 131
  - Least common multiple 136

Part II: Shape 141

13 **Triangles** 143
  - It all adds up to 180 143
  - Area 145
  - Centers 148
  - Lurking equilateral triangles 151

14 Pythagoras and Fermat   155

    The Pythagorean Theorem   155
    Absolute value of complex numbers   158
    Pythagorean triples   159
    Fermat's Last Theorem   162

15 Circles   165

    A precise definition   165
    An equation   166
    Triangles right inside   167
    Ptolemy's Theorem   167
    Packing   168
    Kissing circles   170
    Pascal's Hexagon Theorem   173

16 The Platonic Solids   175

    Polyhedra   175
    Euler's polyhedral formula   178
    Is that all there is?   183
    Archimedian solids   187

17 Fractals   190

    Sierpinski's triangle   190
    Between dimensions   192
    Box counting   194
    The dimension of Sierpinski's triangle   198
    Pascal and Sierpinski   201
    The Koch snowflake   202

18 Hyperbolic Geometry   204
  Euclid's Postulates   204
  What is a line?   207
  An entire plane inside a disk   210
  Implications   211

Part III: Uncertainty   215

19 Nontransitive Dice   217
  A game of two dice   218
  A challenger   219
  Triumph of the underdog   220
  Further examples   220

20 Medical Probability   223
  Conditional probability*   226

21 Chaos   230
  Functions   231
  Iterating the logistic map   233
  From order to chaos   237
  The Collatz $3x+1$ problem   242

22 Social Choice and Arrow's Theorem   245
  Two-party elections   245
  Elections with more than two candidates   249
  The independence of irrelevant alternatives   255

23 Newcomb's Paradox 259

    Newcomb's game 259

    Don't leave money on the table! 262

    Greed doesn't pay 263

    Conflict and resolution 265

    Computer as the Chooser 267

Further Reading 269

Index 271

Preface

Joy

JOYFUL, BEAUTIFUL MATHEMATICS. People are familiar with the masterpieces of various disciplines. In art, we have the *Mona Lisa*, in theater *Hamlet*, in biology the role of DNA in heredity, in archaeology the deciphering of hieroglyphics via the Rosetta stone, and in physics the equation $E = mc^2$. The masterpieces of mathematics are more elusive, and my goal is to share with you some personal favorites.

Art museums own vast collections but are able to display only a fraction of their inventory. Likewise, as the "curator" for this volume, I had to make some tough choices about which masterpieces to present.

I am not limited to exhibiting only one mathematical gem, but if I were, this would be my choice: the proof that there are infinitely many primes. This one example epitomizes my priorities in choosing the topics for this *Companion*:

- It is not well known to nonmathematicians. Readers may know what a prime number is but may not have considered the question: How many primes are there?

- It highlights the idea of *proof* and, in particular, the technique of *proof by contradiction*.

- It does not rely on college-level mathematics. We can do all our work with the tools a high school

For some, describing mathematics as "joyful" and "beautiful" might seem incongruous, but we should not confuse wonderful mathematics with laborious arithmetic any more than we should equate reading great literature with rote learning of spelling.

The proof that there are infinitely many primes is presented in Chapter 1.

student would typically encounter.

- There is an element of surprise. The answer is not obvious. While it is easy to see that there are infinitely many even numbers or perfect squares, there is no clear-cut pattern to the sequence of primes. It is also amazing that a short bit of reasoning leads to the inescapable conclusion that the primes go on forever.

- It is linked to a practical application: in this case, cryptography.

While the various topics explored in this *Companion* don't necessarily exhibit all these features, every chapter contains mathematical marvels that will surprise and intrigue the reader.

IN 1940, THE BRITISH MATHEMATICIAN Godfrey H. Hardy published *A Mathematician's Apology* as a personal justification for spending his life in the study of abstract mathematics. In his *Apology*, Hardy explains the joy and fulfillment he experienced. But trying to explain the joy of mathematics is something like trying to explain the joy of swimming. Until one can float a bit and splash around in refreshing water, it is difficult to understand how swimming can be fun.

I fear that too many people's mathematics education is devoid of joy. Imagine if children's reading education focused primarily on spelling and punctuation, but not on delights such as *Harry Potter* or creating stories of one's own; that approach would hardly instill students with a love of literature.

Here is a caricature of how some people might think of their mathematics education:

- In elementary school I had ten oranges, but someone took three of them away. Why would they do that? I would have shared.

- In middle school I found common denominators. And percentages.

- In high school I learned the quadratic formula—and I can still repeat it by heart—but I don't know why I should care.

$$\frac{-b \pm \sqrt{b^2 - 4ac}}{2a}$$

Certainly, mathematics has great practical applications, but it also has profound beauty. Our goal is to share a bit of that beauty.

Overview

MATHEMATICS IS THE STUDY OF NUMBERS AND SHAPES; accordingly, I have chosen these two concepts as themes for the first two parts of this *Companion*.

In *Number* we explore some specific numbers (such as $\sqrt{2}$ and e) as well as sequences of numbers (such as the primes and the Fibonacci numbers). There are plenty of surprises in store for the reader such as how one infinity can be "more infinite" than another and why there are more numbers that begin with the digit 1 than with a 9.

In *Shape* we visit some familiar friends (such as triangles and circles) but also three-dimensional figures (Platonic solids) and shapes that are more than one-dimensional but less than two-dimensional (fractals). Many surprises await. For example, it is easy to see how to completely tile a floor with squares or with regular hexagons, but it's also "possible" with regular pentagons. Surprised? Intrigued? That's my hope.

A tiling with regular hexagons.

We conclude with a part entitled *Uncertainty* for ideas that are random, unpredictable, and counterintuitive. How is it that a highly accurate medical test might give mostly incorrect results? Do rankings have any meaning? What is the "best" way to elect public officials when more than two candidates are vying for office? As before, surprises await.

Each chapter stands on its own, and you are welcome to read them in any order. The difficulty of the material varies, and it is fine to set aside more challenging parts to revisit later.

A few chapters do refer to material in earlier chapters, but this interdependence is mild.

How to read a mathematics book

TAKE YOUR TIME. The chapters in this book are short but it will take time and effort to grasp the ideas. I often give some calculations or algebra to back up various points; readers can better understand what's going on by working through the steps with a pencil and a pad of paper. It may be necessary to reread material a few times before it truly makes sense.

If possible, do not read this alone. Find a partner and talk the ideas out to each other. This will force you to repeat the material aloud in a way that makes sense to your partner. This will help you truly own the concepts.

The chapters are designed so that the more complicated ideas are toward the end. It's fine to get part way through a chapter and decide "that's enough" before moving on to another chapter.

About the cover

On the cover of this book you can find this equation:

$$\left(x^2 + y^2 - 1\right)^3 = x^2 y^3. \qquad (*)$$

What pairs of numbers (x, y) satisfy this equation? For example, when $x = 1$ and $y = 0$ both sides of $(*)$ evaluate to the same number, namely 0. Likewise, when $x = -1$ and $y = 1$ both sides of $(*)$ evaluate to 1. In other words, both $(1, 0)$ and $(-1, 1)$ are solutions to this equation. Notice that $(0, 0)$ is not a solution.

There are infinitely many solutions to this equation including this:

$$x = 0.70711\ldots \quad \text{and} \quad y = -0.41401\ldots.$$

With these values, both sides of $(*)$ evaluate to $-0.03548\ldots$.

Although there are infinitely many solutions to $(*)$, we can visualize them all by plotting the graph of this equation. This means placing dots in the plane at coordinates (x, y) for all the solutions to $(*)$. When we do, an image appears and it is the heart-shaped curve proudly displayed on the cover of this book.

Don't you just love mathematics? By the time you finish this book, you certainly will!

Acknowledgments

I WANT TO THANK several people who gave me excellent feedback and helpful suggestions in the preparation of this book. This includes Mordechai Levy-Eichel, Joshua Minkin, Yoni Nadiv, Amy Scheinerman, Daniel Scheinerman, Jonah Scheinerman, Leora Scheinerman, Naomi Scheinerman, and Rachel Scheinerman, who commented on early drafts and made helpful suggestions.

An extra thanks to Danny for his idea for the title of this book and to Jonah for his drawing of the spyglass on page 67.

In the course of evaluating this book for publication, I received excellent feedback from reviewers, most of whom were anonymous, but some were happy to share their identities (and kind comments) with me. Thank you to Christoph Börgers, Anna Lachowska, and Jayadev Athreya for your feedback and enthusiasm.

I also want to thank Art Benjamin for the Texas Hold 'Em material in Chapter 19. This example can be found as an exercise in Stewart N. Ethier, *The Doctrine of Chances: Probabilistic Aspects of Gambling*, Spring 2010.

Finally, many thanks for all the help I received

from Yale University Press. First of all, thanks to my editor Joe Calamia for his enthusiasm, his many helpful suggestions, and for answering my ceaseless questions. Thanks also to Ann-Marie Imbornoni for close support in the preparation of the final version, to Liz Casey for her meticulous copy editing, to Eva Skewes for administrative support, to Sonia Shannon for the overall design, and to Thomas Starr for the delightful cover.

Prelude: Theorem and Proof

> "Beauty is truth, truth beauty," – that is all
> Ye know on earth, and all ye need to know.
>
> John Keats, *Ode on a Grecian Urn*

> Beauty is the first test: there is no permanent place in the world for ugly mathematics.
>
> G. H. Hardy, *A Mathematician's Apology*

WHAT DO WE MEAN when we say that something is *true*? In the sciences, truth is demonstrated through observations, often in the form of an experiment. We know that planets travel around the sun in elliptical orbits because Johannes Kepler painstakingly pored over Tycho Brahe's measurements to reach this conclusion. We know that the speed of light in a vacuum is a constant, again because of repeated direct measurements.

It turns out that planetary orbits are not exactly elliptical because their gravitational fields interact with each other, and not just with the gravity of the sun. And we don't know for sure that the speed of light in our galaxy is the same as in, say, the Andromeda galaxy because we've not ventured there to take measurements.

In science, *truth* is not absolute; it is an ever improving sequence of approximations. We think the earth is flat, and for most daily matters, that's a perfectly good approximation. However, as soon as we want to travel an appreciable distance from home, that approximation fails us. A much better model is that the earth is a sphere, and that works great. Great, that is, until we travel on a global scale, and then a better model is that the earth is an oblate spheroid: the circumference around the equator is a bit greater than the circumference through the poles. This shape is predicted by theory and then verified by measurement.

> Even the assertion that the earth is an oblate spheroid is not quite right; it does not take mountains and valleys into account.

In mathematics, on the other hand, *truth* is absolute. When we assert that the sum of two odd integers is even, we mean that this is always true, 100% guaranteed. How do we know? Because we can *prove* it.

A MATHEMATICAL PROOF RESULTS IN COMPLETE CERTAINTY. Other realms use the word *proof*. For example, DNA evidence can be used as proof to establish guilt or innocence. But this is not absolute. The tests are highly accurate, but not perfect. DNA collected from a crime scene might be contaminated. The perpetrator might have an identical twin. Finding DNA at a crime scene does not tell us what the accused person did at that location; only that his/her DNA was found there.

In mathematics, the standards for truth and its verification are absolute. True mathematical assertions are called *theorems*. Here's a simple example: *The sum of two odd integers is even*. For example, 3 is odd and 11 is odd, so their sum $3 + 11 = 14$ is even. The assertion that the sum of two odds is even holds absolutely and without exception.

How do we know? We can add pairs of odd numbers over and over again, and in every case observe an even result. That's how science works, but not mathematics. We are absolutely certain that this theorem is true because we can give a *proof*.

To illustrate, we give the proof here. First, we need

to be precise about what *odd* and *even* mean. Here are the definitions:

- An integer X is called *odd* if we can find an integer a such that $X = 2a + 1$. For example, 13 is odd because we can express 13 as $2 \times 6 + 1$.

- An integer X is called *even* if we can find an integer a such that $X = 2a$. This is a fancy way of saying that an even integer is the result of doubling an integer. For example, 20 is even because $20 = 2 \times 10$.

With these definitions in place, we prove the theorem that the sum of two odd integers is even.

Proof. Let X and Y be odd integers. That means that $X = 2a + 1$ and $Y = 2b + 1$ where a and b are integers. The sum of X and Y can be expressed algebraically as follows:

$$X + Y = (2a + 1) + (2b + 1) = 2a + 2b + 2 = 2(a + b + 1).$$

Notice that $X + Y$ is 2 times an integer, namely 2 times $a + b + 1$. Therefore $X + Y$ is even. □

Notice that a proof is not a bunch of equations. It's an essay composed of complete sentences that takes us from assumptions (that X and Y are odd integers) step by step to an inescapable conclusion (that $X + Y$ is even).

Creating a proof is challenging but much more enjoyable than reading someone else's, so we invite you to try this: Prove that when two odd integers are multiplied together, the result is also odd. Try this on your own, and then compare with our solution on the next page.

Hint: The first sentence of your proof should be "Let X and Y be odd integers." The last sentence of your proof should be "Therefore XY is odd."

Other theorems of mathematics are more interesting and their proofs are more complicated, but the goal is the same: to establish a mathematical fact with 100% certainty.

In summary:

> A *theorem* is a statement about mathematics that is shown to be incontrovertibly true by a *proof*.

The last words

WHAT ARE THOSE THREE LITTLE WORDS that a mathematician longs to hear?

Certainly, we appreciate "I love you" as much as anyone, but in this case the magic phrase is *quod erat demonstrandum*. This Latin phrase roughly translates to "which is what we were to prove" and traditionally was written at the end of a mathematical proof. However, few people wrote these words in their entirety preferring, instead, to use just the initials QED. Sadly, even this abbreviation is out of style, and the fashion these days is to use a symbol, such as a little square □, to mark the end of a proof.

The product of odd integers is odd.

Proof. Let X and Y be odd integers. That means that $X = 2a + 1$ and $Y = 2b + 1$ where a and b are integers. The product of X and Y can be expressed algebraically as follows:

$$XY = (2a+1)(2b+1) = 4ab + 2a + 2b + 1 = 2(2ab + a + b) + 1.$$

Notice that XY can be expressed in the form $2c + 1$ where $c = 2ab + a + b$ is an integer. Therefore XY is odd. □

PART I: NUMBER

1
Prime Numbers

THE PHYSICIST RICHARD FEYNMAN believed that if humanity were to be faced with the loss of all scientific knowledge but was able to pass on just one sentence about science to this postapocalyptic world, that sentence should describe how matter is composed of atoms. In that spirit, if we could pass on only one bit of mathematics to the next generation, it should be the solution to the problem: How many prime numbers are there?

> Feynman's sentence: "All things are made of atoms— little particles that move around in perpetual motion, attracting each other when they are a little distance apart, but repelling upon being squeezed into one another."

The integers

MATHEMATICAL THOUGHT BEGINS with counting. The numbers we use to count are familiar: 1, 2, 3, and so forth. The absence of objects to count—and the need to give a number to that absence—leads us to the number 0. When we add or multiply these counting numbers the result is always another counting number. But subtraction gives us some trouble. All is well when we subtract three from five, $5 - 3$, but if we want to subtract the other way around, that is $3 - 5$, the result is not a counting number. We fix this deficit by introducing the negative numbers -1, -2, -3, and so on.

Collectively, these positive and negative whole numbers together with zero are known as the *in-*

tegers. Mathematicians use a stylized capital Z to denote the collection of all integers:

$$\mathbb{Z} = \{\ldots, -4, -3, -2, -1, 0, 1, 2, 3, 4, \ldots\}.$$

DIVISION IS A PROBLEM for the integers. While we may add, subtract, or multiply two integers and be assured that the result is an integer, division of two integers might or might not be an integer.

Given two positive integers a and b, we say that a is *divisible* by b provided that $a \div b$ is also an integer. We also say that b is a *factor* of a and that b is a *divisor* of a.

For example, 24 is divisible by 6 (because $24 \div 6$ is an integer) but 24 is not divisible by 7 (because $24 \div 7$ is not). Every positive integer is divisible by itself: if a is a positive integer, then $a \div a = 1$, which, of course, is an integer. Every positive integer is divisible by 1 because if a is a positive integer, then $a \div 1 = a$.

A positive integer is called a *prime* if it has precisely two positive divisors: 1 and itself.

For example, 17 is a prime because its only positive divisors are 1 and 17. Likewise, 2 is prime.

On the other hand, 18 is not prime because, in addition to being divisible by 1 and itself, it is also divisible by 2, 3, 6, and 9. A number such as 18 is called *composite*. Specifically a positive integer is called *composite* provided it has a positive divisor in addition to 1 and itself.

This classification of numbers as either prime or composite accounts for all the positive integers except one: 1. We call 1 a *unit*. Just as some people are bothered by the fact that Pluto is not considered a planet, others are "offended" by the fact that 1 is not deemed a prime. We'll explain why 1 gets its own category in a bit.

Summarizing, we have these three categories of positive integers:

> It's a bit strange to create a name (*unit*) for a category of numbers when there's only one number (1) in that category. In fact, the term *unit* has a broader context in advanced mathematics that, when applied just to the positive integers, reduces to the single number 1.

- *units* with exactly one positive divisor,
- *primes* with exactly two positive divisors, and
- *composites* with three or more positive divisors.

Notice that 1 is the only positive integer that is a unit but there are infinitely many composite numbers: the numbers 4, 6, 8, 10, 12, and so on are all composite (and there are many more).

How many primes are there?

Factoring

FACTORING IS THE PROCESS of expressing a positive integer as the answer to a multiplication problem. Consider the number 84. We can factor 84 in several different ways including

$2 \times 42, \quad 3 \times 28, \quad 12 \times 7, \quad 2 \times 6 \times 7, \quad \text{and} \quad 21 \times 4.$

The ultimate way to factor 84 is to refine all of the terms to primes, like this: $84 = 2 \times 2 \times 3 \times 7$. We cannot break down any of these factors into smaller bits because every term is a prime. Of course, we could include additional factors of 1 like this

$$84 = 1 \times 1 \times 2 \times 2 \times 3 \times 7,$$

but these extra terms clutter rather than simplify the expression, and they do not reduce any of the terms into smaller factors.

Let's consider another example: 120. We can start by factoring 120 into 12×10 and then break 12 down as $2 \times 2 \times 3$ and 10 as 2×5 giving

$$120 = (2 \times 2 \times 3) \times (2 \times 5). \quad (A)$$

Alternatively, we can start with $120 = 4 \times 30$ and then observe that $4 = 2 \times 2$ and $30 = 2 \times 3 \times 5$. Combining these gives

$$120 = (2 \times 2) \times (2 \times 3 \times 5). \quad (B)$$

> This is the reason we craft the definition of *prime* to exclude the integer 1. We think of the primes as the irreducible nuggets that we use to build any positive integer through multiplication. The number 1 is not useful in this regard.

What's important to notice is that the prime factors in expressions (A) and (B) are the same except for the order in which they appear. This is shown graphically in the figure.

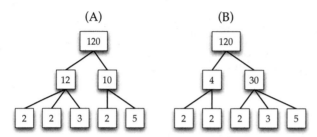

Any method that reduces 120 to a product of primes gives exactly the same result.

This *unique factorization* property is established in the following theorem.

Theorem (Fundamental Theorem of Arithmetic). *Every positive integer can be factored into primes and that factorization is unique (up to the order of the factors).*

(A few words of explanation are in order. For a number such as 30 the meaning is clear. We can factor 30 as $2 \times 3 \times 5$ or $5 \times 3 \times 2$, but these are the same—only the order of the terms differs. For a prime number such as 13, the only prime factorization of 13 is into the single term 13 by itself. But what about 1? By convention, an *empty product* evaluates to 1; that is, a multiplication problem with no terms is declared to have the value 1.)

By combining primes together we build all the positive integers. Prime numbers are the atoms of multiplication.

A *theorem* is a statement about mathematics that can be incontrovertibly proved.

A theorem is wholly different from a *scientific theory*, which is a model or explanation of some facet of the physical world that is backed up by experimental evidence. And this is different again from a *mathematical theory*; that's a body of definitions and theorems in a particular topic.

We do not present the proof of the Fundamental Theorem of Arithmetic. That can be found in most books on number theory: the mathematical study of the properties of integers. Raising a number to the 0th power is an example of an empty product. By definition, 10^n is the result of multiplying 10 with itself n times. In the case $n = 0$, we have that 10^0 is 1: the result of multiplying no terms together!

How many?

LET US RETURN TO THE QUESTION: How many primes are there? Here's the answer:

Theorem. *There are infinitely many primes numbers.*

This result is attributed to Euclid. Its proof is a mathematical gem. We cannot *prove* this theorem by collecting data. We might observe that primes appear frequently; the first several primes are

2, 3, 5, 7, 11, 13, 17, 19, 23, 29, 31, 37, 41, 43, 47, 53, 59, 61, and 67.

But as one looks deeper into the sequence of primes, the gaps become wider. Just looking at the list of primes above, the largest gap between consecutive primes is 6 (for example, between 53 and 59). But 89 and 97 are primes that are 8 apart and all the integers between them are composite. Likewise, 139 and 149 are consecutive primes that are 10 apart. Were we to look further, there are increasingly large gaps between consecutive primes. This may suggest that eventually the primes "die out." In fact, although primes become rarer as we explore the world of larger and larger integers, the list of primes is endless. But, in order to be convinced, we need proof.

The key idea is to ask: What if?

WHAT IF THERE WERE ONLY FINITELY MANY PRIMES? If we can show that *supposing* that there are only finitely many primes leads us to an absurd conclusion, then that supposition must be false. And, like Sherlock Holmes, once we have eliminated the impossible (finitely many primes) what remains must be the truth: there are infinitely many.

Here's a high-level overview of the argument.

1. Assume there only finitely many primes.

2. Show that this supposition leads to an impossible outcome.

3. Conclude that since the supposition leads to a logical impossibility, it must be wrong.

4. Therefore, there are infinitely many primes.

> This process is like trapping an accused criminal in a lie:
> "You say you were at home on the night in question, Mr. Nogoodnick?"
> "Yes."
> "What were you doing?"
> "Watching television."
> "Are you aware there was a power failure that evening?"
> "Uh …"
> Evidently, Mr. Nogoodnick was not at home watching television!

Let's get to work. We suppose that there are only finitely many primes and see where that supposition leads us.

If there are but finitely many primes, then there is a *largest* prime P—the very last prime in the sequence of primes. The entire list of primes looks like this:

$$2, 3, 5, 7, 11, 13, \ldots, P.$$

Next, we multiply all these primes together and add one to the result. Let's call that huge number N:

$$N = (2 \times 3 \times 5 \times 7 \times 11 \times 13 \times \cdots \times P) + 1.$$

Let's imagine for a moment that 13 is the last prime. Were that the case, then N would equal

$$(2 \times 3 \times 5 \times 7 \times 11 \times 13) + 1,$$

which is 30,031.

Is N prime? According to our supposition, the answer must be no because N is larger than P, the last prime. *Therefore N is composite and must be divisible by primes.* Here is where we run into trouble.

We know that N has a prime factor; might that prime factor be 2? We claim N cannot be divisible by 2. Look at the expression for N and notice that the part in parentheses is even because it includes the factor 2:

$$N = (\underline{2} \times 3 \times 5 \times 7 \times 11 \times 13 \times \cdots \times P) + 1.$$

Thus N is one larger than a (huge) even number; in other words, N is odd and therefore *not* divisible by 2.

That's OK. We know that N has a prime factor; no worries that 2 happens not to be such a factor. What about 3? Look again at the term in parentheses and notice that it is a multiple of 3:

$$N = (2 \times \underline{3} \times 5 \times 7 \times 11 \times 13 \times \cdots \times P) + 1.$$

Thus N is one larger than some (huge) multiple of 3. That means if we were to try to calculate $N \div 3$, the division would leave a remainder of 1. Therefore N is not divisible by 3.

See where we're heading? Let's try the next prime, 5. We claim that N is not divisible by 5 because it is one larger than a multiple of 5:

$$N = (2 \times 3 \times \underline{5} \times 7 \times 11 \times 13 \times \cdots \times P) + 1.$$

In a similar manner, we conclude that N is not divisible by 7 or 11 or 13 or ... or any prime!

What have we learned? Our supposition that there are only finitely many primes leads us to two conclusions:

- N is divisible by a prime.

- N is not divisible by any prime.

This cannot be! The only way out of this pickle is to realize that the supposition that there are finitely many primes is false. Therefore there are infinitely many primes!

A constructive approach

THE PROOF WE JUST PRESENTED is known as a *proof by contradiction*. In this proof *we suppose the opposite* of what we want to demonstrate is true, show that this leads to an impossible situation, and conclude that the supposition we made must have been false, and therefore the statement we sought to prove is true. This is some fantastically fancy fallacy footwork.

But there is another way to think about the proof: in its essence, it is a prime-making machine. We feed a handful of primes into the machine and, *voilà*, out come some new primes. Here's how the machine works.

Let's grab a half-dozen primes: 2, 3, 5, 7, 11, and 13. We multiply all these primes together and add one to make a new number:

$$(2 \times 3 \times 5 \times 7 \times 11 \times 13) + 1 = 30{,}031.$$

We note that 30,031 is not divisible by 2; this is easy to tell because its last digit is odd. It's also not divisible by 3 (because it is one larger than 2 × 3 × 5 × 7 × 11 × 13, which is a multiple of 3). Likewise, it is not divisible by any of 5, 7, 11, or 13. So either it is itself a prime that was not part of our original list, or it factors into primes that were not on the original list. It so happens that 30,031 is composite and factors as 59 × 509; 59 and 509 are not on the original list.

We can now take these new primes, together with the original half-dozen and produce a new number:

$$(2 \times 3 \times 5 \times 7 \times 11 \times 13 \times 59 \times 509) + 1,$$

which equals 901,830,931, which happens to be prime.

We can throw this new prime into the growing list and crank out a new number that either is itself yet another prime or factors into other additional primes. There's no limit to how often we may repeat this process to produce more and more primes.

There are sophisticated methods to determine whether or not a number is prime; these methods can complete their work rapidly on an ordinary computer.

A different proof

THERE ARE MANY PROOFS THAT THERE ARE INFINITELY MANY PRIMES; here is another.

As in the first proof, we suppose that there are only finitely many primes and show that this reasoning leads to a contradiction. We imagine that the largest prime is P so the entire list of primes is

$$2, 3, 5, 7, 11, 13, \ldots, P.$$

Let N be the result of multiplying all these numbers together:

$$N = 2 \times 3 \times 5 \times 7 \times \cdots \times P.$$

Let's think about all the numbers from 1 to N inclusive. Of these, each of them, except for 1, is divisible by one or more of these primes; after all, every number (except 1) is divisible by some prime.

How many of the numbers from 1 to N are divisible by 2? Clearly, it's half of them (the even numbers). Let's cross them out leaving just the odd numbers remaining:

$1, 3, 5, 7, 9, 11, 13, 15, 17, 19, 21, 23, 25, 27, 29, 31, 33, 35, 37, 39, 41, 43, 47, \ldots$

The number of integers between 1 and N that we have not crossed out is $N/2$.

From the numbers remaining, let's strike out the multiples of 3. Here's what remains:

$1, 5, 7, 11, 13, 17, 19, 23, 25, 29, 31, 35, 37, 41, 43, 47, 49, 53, 55, 59, 61, 65, \ldots$

This deletes one third of the remaining numbers. The number of uncrossed numbers is $\frac{2}{3}$ of what we had before, that is, $N \times \frac{1}{2} \times \frac{2}{3}$.

We continue with striking out the multiples of 5, thereby deleting one fifth of the remaining numbers and leaving $N \times \frac{1}{2} \times \frac{2}{3} \times \frac{4}{5}$ of the numbers. Here's what remains so far:

$1, 7, 11, 13, 17, 19, 23, 29, 31, 37, 41, 43, 47, 49, 53, 59, 61, 67, 71, 73, 77, 79, \ldots$

The number of integers that have not been crossed out so far is $N \times \frac{1}{2} \times \frac{2}{3} \times \frac{4}{5}$.

Next we strike out the multiples of 7, leaving $\frac{6}{7}$ of the list intact and so on until finally we delete the multiples of P.

At the end, the number of numbers we have *not* crossed out is

$$N \times \frac{1}{2} \times \frac{2}{3} \times \frac{4}{5} \times \frac{6}{7} \times \frac{10}{11} \times \cdots \times \frac{P-1}{P}. \qquad \text{(C)}$$

Since all the numbers from 1 to N, except for 1, are divisible by some prime, expression (C) should evaluate to 1. Does it? Remember that $N = 2 \times 3 \times 5 \times \cdots \times P$; when we substitute that into expression (C) and spread its factors over the matching denominators, we get

$$\left(2 \times \frac{1}{2}\right) \times \left(3 \times \frac{2}{3}\right) \times \left(5 \times \frac{4}{5}\right) \times \left(7 \times \frac{6}{7}\right) \times \cdots \times \left(P \times \frac{P-1}{P}\right),$$

which equals

$$1 \times 2 \times 4 \times 6 \times \cdots \times (P-1),$$

which is *much* bigger than 1! Expression (C), which must equal 1, is clearly not equal to 1. The "mistake" is in assuming that there are only finitely many primes. Therefore, there must be infinitely many.

Two infamous problems

THERE ARE MANY FASCINATING PROBLEMS ABOUT PRIME NUMBERS; here we'll introduce two of the most notorious.

Although there are infinitely many prime numbers, they become increasingly sparse as we venture to greater and greater numbers. Later (Chapter 7) we consider the average gap between large primes. Nevertheless, primes often appear close together. Except for the primes 2 and 3, the closest two primes can be to one another is a difference of two. Primes that are two apart are known as *twin primes*. The smallest twin prime pair is 3 and 5. Between 1 and 10,000 there are 205 twin prime pairs, the last of which is 9929 and 9931.

The question is: Are there infinitely many twin prime pairs?

To date, we don't know.

HERE IS ANOTHER PROBLEM attributed to Christian Goldbach, a German mathematician who lived 1690–1764, who wondered if even numbers (other than 2) can be expressed as the sum of two primes. For example:

$4 = 2+2 \quad 6 = 3+3 \quad 8 = 3+5 \quad 10 = 3+7$
$12 = 5+7 \quad 14 = 7+7 \quad 16 = 5+11 \quad 18 = 5+13$
$20 = 3+17 \quad 22 = 11+11 \quad 24 = 7+17 \quad 26 = 13+13$

The problem is: Can we continue this forever? Specifically, Goldbach conjectured that every even number (greater than 2) is the sum of two primes.

To date, we don't know.

Application to cryptography

THE STUDY OF PRIME NUMBERS belongs to a branch of mathematics known as *number theory*. Of number theory, the British mathematician Godfrey Harold Hardy wrote

> No one has yet discovered any warlike purpose to be served by the theory of numbers ...

G. H. Hardy, *A Mathematician's Apology* (Cambridge University Press, 1940).

Hardy could not foresee our worldwide network of computers of the fact that the security of this network has come to depend on prime numbers. How so?

Let P and Q be two large prime numbers—say, one hundred digits each. While laborious for a person to multiply, a computer can calculate the product $N = P \times Q$ in a flash. However, if we are only told the answer, N, and asked to work backward to find its two prime factors, we are stymied. No one knows an efficient method to find the factors of such large numbers.

(It is curious that determining whether a number is composite can be accomplished rapidly; finding large factors of a composite number, however, is difficult.)

To illustrate the relative ease of multiplication compared to the opposite operation of factoring, try working these two problems with just pencil and paper. One the one hand, multiply 227×281. If you work carefully, you can find the six-digit answer in a few minutes. On the other hand—and by hand—try to find two three-digit primes whose product is 211,591. That's not so much fun. The answer is on the following page.

Amazingly, this asymmetry—easy to multiply versus difficult to factor—has been exploited to develop secret codes. A *public key cryptosystem* is one in which a person is able to fully disclose a method for putting messages into code without revealing the method for decrypting the secret back into plain language. While the details of this method are beyond our scope, the main idea is that the encryption key uses a composite number N that is the product of two large primes: $N = P \times Q$. The decryption method requires

The term *public key* refers to the fact that disclosing the encryption procedure—the key to the encryption method is public—does not compromise the decryption method. One method for accomplishing this was developed in the 1970s by Ron Rivest, Adi Shamir, and Leonard Adleman; their method bears their initials, RSA.

knowledge of the individual numbers, P and Q. Revealing N does not disclose its factors because it is enormously difficult to calculate those divisors.

Whenever we make an online purchase, public key cryptography springs into action. Before the web browser sends our credit card number to a merchant, it obtains the merchant's public key encryption method. Our browser uses that method to encode the credit card number. An eavesdropper on the internet learns nothing because knowledge of the encryption method does not compromise the decryption method (that only the merchant knows). Once the encrypted message arrives in the merchant's computer, the private decryption method is used to reveal the credit card number just for the intended recipient.

Public key cryptography has military utility, perhaps even to control the deployment of atomic bombs.

Solution to the factoring problem: $211{,}591 = 457 \times 463$.

2
Binary

When in Rome

THE ROMANS ARE OFTEN CRITICIZED for their unwieldy way of writing numbers. Roman numerals are maligned because they make computation painful. There's no nice way to multiply XLVII by DCDXXIV. On the other hand, when written as 47 × 924 the problem is less intimidating. Still, most of us would reach for a calculator to work this out.

But before we dismiss the Roman system as a quaint anachronism, we should acknowledge that its underlying concept—letters stand for numeric values—is quite useful. Indeed, that key aspect of Roman numerals is still in use today but in a new incarnation. Which of the following is more readable?

- Renovating the county's high schools will cost $23000000.

- Renovating the county's high schools will cost $23M.

Of course, I omitted the commas from the first figure deliberately to make it difficult to read (and thereby make the point). But even with the commas, reading "the Pentagon is requesting an additional $19,000,000,000" is more difficult to parse than "an

A classic mathematics T-shirt reads: "There are 10 kinds of people in this world: Those who understand binary and those who don't." When you finish reading this chapter, you'll be in on the joke.

additional $19B." Using letters as abbreviations for numbers is handy.

A purported advantage of the place-value system we use is that it simplifies calculation. But let's consider the effort we go through to multiply two numbers. First, we memorize our "math facts." That is, we commit to memory the addition and multiplication tables (including the dreaded 8×7). Then we learn a lengthy procedure in which we line numbers up neatly in columns, multiply each digit in one number by a digit in the other, keep track of carries, and end with a multiline addition problem.

True, decimal numerals are easier to multiply than their Roman cousins, but it is still taxing. So it is natural to consider if there is another way to write numbers that makes computation easier. What we find is that there are options that make calculation simpler, but at the cost of clarity.

Unary

THE SIMPLEST WAY TO REPRESENT NUMBERS is using *unary*. In this method, we simply make as many marks (we'll use the digit 1) as the number we wish to represent. That is, to write the number three, we make three marks: 111. Addition and multiplication become exceptionally easy. To add three and five, we simply write the two numbers, 111 and 11111, next to each other (with no space between), to find the answer 11111111. Multiplication is also easy. We write one of the numbers vertically and the other horizontally to form a chart like this:

$$
\begin{array}{c|cccc}
 & 1 & 1 & 1 & 1 & 1 \\
\hline
1 & & & & & \\
1 & & & & & \\
1 & & & & & \\
\end{array}
$$

We then complete the innards of the chart placing a mark in every row and every column:

	1	1	1	1	1
1	1	1	1	1	1
1	1	1	1	1	1
1	1	1	1	1	1

Finally, we take all the marks from the inside of the chart and run them together to arrive at the answer: 111111111111111. It is much easier to add and multiply numbers when they are expressed in unary than when they are given as decimal or Roman numerals.

We encourage the reader to develop simple procedures for subtraction and division.

Of course, this computational simplicity comes at a terrible cost of comprehension and utility. We would not want to use this method to calculate 47×924.

A middle ground

NUMBERS EXPRESSED IN BINARY are less easily understood than when written in decimal (or Roman) notation, but calculations become easier. For this reason computers represent numbers internally in binary. To understand how binary notation works, we need to recall some of the details of decimal notation.

Binary is also called base two.

Numbers expressed in decimal notation use ten characters—the digits 0 through 9—written horizontally. The contribution of each digit to the value of the number depends on where in the figure the digit appears. That is, 29 and 92 represent different values because the 2 and the 9 are in different locations in the numbers. Specifically 29 means "two tens and nine." The number 5804 is "five thousands, eight hundreds, no tens, and four." Each place in a decimal numeral stands for a different power of ten. Reading from right to left, the places have the values one, ten, hundred, thousand, ten thousand, and so on. Writing these as 10^0, 10^1, 10^2, 10^3, and so forth, the notation 5804 means

$$\underline{5} \times 10^3 + \underline{8} \times 10^2 + \underline{0} \times 10^1 + \underline{4} \times 10^0.$$

Recall that the exponent gives the number of times we multiply the base, so 10^3 means $10 \times 10 \times 10$. Naturally, 10^1 means 10. By convention, 10^0 means 1. This makes sense because each subsequent power of 10 is ten times larger than the previous.

Each place in a decimal numeral carries a value that is a power of ten. When a decimal figure has more than four digits, it becomes difficult to read. Inserting commas enhances readability.

Binary notation works the same way as decimal, but each place value is a power of two.

In binary notation we use only two digits: 0 and 1. Each place in a binary numeral stands for a different power of two. Starting from the right the place values are one, two, four, eight, sixteen, and so forth. For example, the binary number 10110 means

$$\underline{1} \times 2^4 + \underline{0} \times 2^3 + \underline{1} \times 2^2 + \underline{1} \times 2^1 + \underline{0} \times 2^0,$$

which equals $16 + 4 + 2 = 22$.

Check your understanding: Express 42 in binary and express 11011_{TWO} in ordinary, base ten notation. The answers are on page 26.

To distinguish binary (base two) from decimal (base ten) numerals we may write the word TWO as a subscript, like this: 1101_{TWO}. In case a number with only zeros and ones is meant to be a decimal numeral, we may place the word TEN as a subscript to eliminate any ambiguity: 1101_{TEN}.

Calculation

BINARY NUMBERS ARE MORE DIFFICULT for people to understand than their decimal counterparts; the binary 1011001 is less intuitive than its base-ten equivalent, 89. Binary's advantage comes from its utility in calculation. To begin, instead of dozens of math facts to memorize, we only need these two tables:

+	0	1
0	0	1
1	1	10

and

×	0	1
0	0	0
1	0	1

Note that in the addition table, 10 is the number *two* written, of course, in binary.

Addition of binary numbers is accomplished by the same method as for decimal. Suppose we want to add 10100_{TWO} and 1110_{TWO}. We stack these numbers atop each other right justified:

$$\begin{array}{r} 10100 \\ +1110 \\ \hline \end{array}$$

We now work our way right-to-left adding up the digits in each column and carrying to the next column if needed. In the example, we start by adding the two 0s on the right, giving 0:

$$\begin{array}{r} 10100 \\ +1110 \\ \hline 0 \end{array}$$

Now we work the two's column adding $0 + 1$ (no carry):

$$\begin{array}{r} 10100 \\ +1110 \\ \hline 10 \end{array}$$

The four's column is next. We add $1 + 1$ to give 10 so we write the 0 and carry the 1:

$$\begin{array}{r} 1 \\ 10100 \\ +1110 \\ \hline 010 \end{array}$$

In the eight's column we add $1 + 0 + 1$ to give 10 so, as before, we write the 0 and carry the 1:

$$\begin{array}{r} 11 \\ 10100 \\ +1110 \\ \hline 0010 \end{array}$$

We finish in the sixteen's column. Adding $1 + 1$ to give 10 we write the 0 in the sixteen's column and the 1 in the thirty-two's column:

$$\begin{array}{r} 11 \\ 10100 \\ +1110 \\ \hline 100010 \end{array}$$

We have found that $10100 + 1110 = 100010$. Translating into decimal we have

$10100_{\text{TWO}} = 20$, $1110_{\text{TWO}} = 14$, and $100010_{\text{TWO}} = 34$

and, of course, $20 + 14 = 34$.

MULTIPLICATION OF BINARY NUMBERS is less onerous than base-ten multiplication. The method for binary numbers depends on two ideas: addition of binary numbers (as we just described) and multiplication by powers of two.

For decimal numbers, multiplication by ten is easy; we simply append a zero like this: $23 \times 10 = 230$. Similarly, in binary, multiplication by two is easy; we just append a zero. For example, $1101 \times 10 = 11010$. We can see this is correct by converting to decimal, but it is more illustrative to write 1101 as

$$\underline{1} \times 8 + \underline{1} \times 4 + \underline{0} \times 2 + \underline{1} \times 1$$

and then when we multiply by two we have

$$\underline{1} \times 16 + \underline{1} \times 8 + \underline{0} \times 4 + \underline{1} \times 2.$$

Placing an extra $+ \underline{0} \times 1$ at the end shows that the result is, indeed, 11010.

Multiplying a number by four or eight or any other power of two is just as simple; for example, to multiply by eight (1000_{TWO}) we simply append three zeros.

Multiplication is now a game of shift-and-add. We illustrate this by multiplying 11010 by 1011. To begin, we write the second number like this:

$$1011 = 1000 + 10 + 1.$$

To multiply this by 11010 we write this:

$11010 \times 1011 = 11010 \times (1000 + 10 + 1)$
$= (11010 \times 1000) + (11010 \times 10) + (11010 \times 1)$
$= 11010\underline{000} + 11010\underline{0} + 11010$

where the underlined zeros are the digits we appended to 11010. We can also write this in traditional multiplication form like this:

$$\begin{array}{r} 11010 \\ \times \quad 1011 \\ \hline 11010 \\ 11010 \\ +\quad 11010 \\ \hline \end{array}$$

Finally, we add these terms to give us the answer:

$$\begin{array}{r} 11010 \\ \times \quad 1011 \\ \hline 11010 \\ 11010 \\ +\quad 11010 \\ \hline 100011110 \end{array}$$

Let's convert to decimal to check our work:

$$11010_{TWO} = 16 + 8 + 2 = 26,$$
$$1011_{TWO} = 8 + 2 + 1 = 11, \quad \text{and}$$
$$100011110_{TWO} = 256 + 16 + 8 + 4 + 2 = 286.$$

And, indeed, $26 \times 11 = 286$.

Extensions

BASE TEN NOTATION IS USED for numbers other than integers; the one's place need not hold the rightmost digit. By placing a decimal point, additional places—each signifying one-tenth the value of its predecessor—give us the ability to express fractional values. When we write 34.27 we are writing an abbreviation for

$$3 \times 10 + 4 \times 1 + 2 \times \frac{1}{10} + 7 \times \frac{1}{100}.$$

Base two numbers can also be used to express fractional values. The places to the right of the decimal point each carry a value that is half of its neighbor to the left. For example, 101.011_{TWO} means

$$1 \times 4 + 0 \times 2 + 1 \times 1 + 0 \times \frac{1}{2} + 1 \times \frac{1}{4} + 1 \times \frac{1}{8}.$$

An esoteric way to write one half is 0.1_{TWO}!

> When working in base two, we might not want to call the separator a *decimal* point. Alternative choices are *binary point* or, generically, *radix point*.

TWO AND TEN ARE NOT THE ONLY choices for bases. Base three is known as *ternary*. In ternary, we use the digits 0, 1, and 2; each place in a base three numeral represents a different power of three. For example, 1102_{THREE} equals

$$1 \times 27 + 1 \times 9 + 0 \times 3 + 2 \times 1,$$

which is 38 in base ten.

The places to the right of the decimal point in a ternary number are each one-third the value of its predecessor. Thus,

$$2.102_{\text{THREE}} = 2 \times 1 + 1 \times \frac{1}{3} + 0 \times \frac{1}{9} + 2 \times \frac{1}{27}.$$

> Base sixteen—also known as *hexadecimal*—is used by computer programmers. Just as base ten figures are composed of ten different symbols (the digits 0 through 9) we need sixteen different digits for hexadecimal numerals. Rather than inventing six more characters, the convention is to use the letters A through F to represent the digits ten through fifteen.

Solution to the questions on page 22: Writing 42 as $32 + 8 + 2$, we see that it equals 101010_{TWO}. On the other hand, 11011_{TWO} equals $16 + 8 + 2 + 1$, which is 27.

3
0.999999999999...

Unquestionably, the simplest way to write the number *one* is like this: 1. You may have also encountered the fact that the endlessly repeating decimal 0.9999... is an alternative way to write the same number. In this chapter we take a careful look at this.

The meaning of decimals

The base ten number system is comfortable and works extremely well—most of the time. For whole numbers, the system is exact. The notation 235 is a shorthand for

two hundreds, three tens, and five ones

or in mathematical notation

$$235 = 2 \times 100 + 3 \times 10 + 5 \times 1.$$

For some fractional quantities decimal notation is completely effective. Think about the number 3/4. In decimal notation this may also be written as 0.75. The notation 0.75 means this:

$$7 \times \frac{1}{10} + 5 \times \frac{1}{100} = \frac{7}{10} + \frac{5}{100} = \frac{75}{100}.$$

The decimal 0.75 exactly equals 3/4.

It is good practice to include a leading 0 for a decimal number between 0 and 1; the, perhaps redundant, 0 alerts us to the decimal point. If we see an unadorned .75 it's possible to miss the dot.

However, when we try to write 2/7 in decimal notation, things get ugly. If we punch $2 \div 7$ into a calculator we get an unpleasant 0.28571429, and this is only an approximation; it does not exactly equal 2/7.

A number such as 3/8 can be written exactly as a decimal because we rewrite 3/8 as 375/1000; the denominator is a power of 10. But we can't find an integer A for which

$$\frac{2}{7} = \frac{A}{10^n}$$

because that would imply $2 \times 10^n = 7 \times A$; there's no integer A that can satisfy this equation because no matter what whole number A you choose, the left-hand side is not divisible by 7 and the right-hand side is. Expressing 2/7 exactly as a decimal is not possible. Unless ...

Decimal numbers with infinitely many digits

FOR MATHEMATICIANS, A DECIMAL NUMBER with infinitely many digits has a subtle meaning, and the point of this chapter is to understand that interpretation. Let's return to the main point of this chapter: what does 0.999999... mean and why is it equal to 1?

To begin, instead of thinking of 0.999999... as a single number, let's think of it as a sequence of numbers in which we append one more 9 to the end from one step to the next. That is, the sequence is this:

$$0.9 \quad 0.99 \quad 0.999 \quad 0.9999 \quad \cdots \qquad (*)$$

ad infinitum. Notice that the terms of sequence $(*)$ get successively larger. Not by much, but each term is bigger than the one that came before.

We're going to demonstrate two facts:

1. All terms in the increasing sequence $(*)$ are less than 1.

2. However, if x is any number less than 1, then after some point, the terms in sequence $(*)$ are larger than x.

To see both points, we rewrite sequence $(*)$ as fractions, like this:
$$\frac{9}{10} \quad \frac{99}{100} \quad \frac{999}{1000} \quad \frac{9999}{10000} \quad \ldots$$

Here's a more compact way to rewrite these fractions. Notice that the denominators are just powers of ten: 10^1, 10^2, 10^3, and so on. Each numerator is just one less than the denominator. So the same sequence can be rewritten yet again this way:
$$\frac{10^1-1}{10^1} \quad \frac{10^2-1}{10^2} \quad \frac{10^3-1}{10^3} \quad \frac{10^4-1}{10^4} \quad \ldots$$

Written this way, it's easy to see that then n^{th} term of sequence $(*)$ is
$$\frac{10^n - 1}{10^n}.$$

Seeing the terms this way, it's clear that every term of sequence $(*)$ is less than 1 because every numerator is smaller than its corresponding denominator.

Now we show the second point: if x is any number less than 1, the terms in sequence $(*)$ eventually surpass x.

Because x is smaller than 1, we know that $1 - x$ is positive. If x happens to be very close to 1, then $1 - x$ is tiny, but still positive. Let's multiply $1 - x$ by a power of 10:
$$10^n \cdot (1 - x).$$

Since $1 - x$ is positive, if 10^n is big enough, the result will be greater than 1:
$$10^n \cdot (1 - x) > 1.$$

Expand the left side
$$10^n - 10^n x > 1,$$

move the 1 to the left and the $10^n x$ to the right

$$10^n - 1 > 10^n x,$$

and divide through by 10^n to give

$$\frac{10^n - 1}{10^n} > x.$$

Let's review what we know. On the one hand, the members of the increasing sequence 0.9, 0.99, 0.999, and so forth are all less than 1. On the other hand, if x is any number less than 1, this sequence eventually grows larger than x (and keeps getting larger, moving farther away from x).

This sequence gets inexorably closer and closer to 1. We say that the sequence *converges* to 1. Alternatively, we say that the *limit* of the sequence is 1.

The meaning of a decimal number with only finitely many digits, such as 0.529, is the sum of so many tenths, hundredths, thousandths, and so forth, like this:

$$0.529 = \frac{5}{10} + \frac{2}{100} + \frac{9}{1000} = \frac{529}{1000}.$$

Unfortunately, the language of decimal numbers with only finitely many digits is not rich enough to express numbers such as 2/7. So we need to enrich our vocabulary.

The value of a decimal number with infinitely many terms is the limit of the sequence formed by appending one digit at a time. This is a lot more complicated but gives us the ability to express all numbers using decimals.

Let's be reckless

THINKING OF AN INFINITE DECIMAL NUMBER as the limit of a sequence requires a great deal of effort. Let's see if we can come up with something simpler.

Let's think about our friend 0.999999... like this. Let

$$X = 0.999999\ldots \qquad (A)$$

Multiply both sides by 10 to get this:

$$10X = 9.999999\ldots \qquad (B)$$

Now subtract line (A) from line (B):

$$10X = 9.999999\ldots$$
$$X = 0.999999\ldots$$
$$\Rightarrow \quad 9X = 9.000000\ldots$$

Finally, divide both sides by 9 and conclude that $X = 1$. Done. Simple.

This trick can be used for any repeating decimal. For example, let

$$Y = 0.27272727\ldots \qquad (C)$$

Multiply both sides by 100 (to get the digits to line up)

$$100Y = 27.272727\ldots \qquad (D)$$

and subtract (C) from (D):

$$100Y = 27.272727\ldots$$
$$Y = 0.272727\ldots$$
$$\Rightarrow \quad 99Y = 27.000000\ldots,$$

which gives $Y = 27/99$, which simplifies to $3/11$.

See! We didn't have to bother with "sequences" and "limits." But here's what can happen if you just throw around infinite sequences carelessly. Consider this sum:

$$Z = 1 + 2 + 4 + 8 + 16 + 32 + \cdots \qquad (E)$$

Multiply through by 2

$$2Z = 2 + 4 + 8 + 16 + 32 + \cdots \qquad (F)$$

> Check your own understanding by expressing 0.123123123123... as a fraction. The answer is on the next page.

and subtract (E) from (F)

$$Z = 1 + 2 + 4 + 8 + 16 + 32 + \cdots$$
$$2Z = \phantom{1 + {}} 2 + 4 + 8 + 16 + 32 + \cdots$$

$$\Rightarrow \quad -Z = 1,$$

which implies that $Z = -1$. This, of course, is nonsense.

WHAT WENT WRONG? WE GOT RECKLESS. The analysis that correctly evaluated $0.999999\ldots$ and $0.272727\ldots$ failed us when we applied it to $1 + 2 + 4 + 8 + 16 + \cdots$. In all three cases, the notation represents a sum of infinitely many terms. The first can be written like this:

$$0.9 + 0.09 + 0.009 + 0.0009 + \cdots$$

How is that different from $1 + 2 + 4 + 8 + 16 + \cdots$? The difference is convergence. Without a careful understanding of convergence, we could be led to think that the sum of infinitely many positive numbers has a negative value! The manipulations we made in equations (A) and (B) [and also (C) and (D)] are mathematically valid because the sequences converge.

Solution to question on the preceding page: Let $x = 0.123123123\ldots$. Then $1000x = 123.123123\ldots$. Subtracting gives $999x = 123$ and so $x = \frac{123}{999} = \frac{41}{333}$.

$\frac{4}{\sqrt{2}}$

BEFORE THE ORCHESTRA BEGINS TO PLAY the musicians tune up to ensure that all their notes harmonize sweetly. Achieving this is mathematically impossible. Let's see why.

Rational numbers

THE INTEGERS PLAY NICELY with the basic operations of addition, subtraction, and multiplication. Given two integers, the result of any of these operations is also an integer. However, the division of two integers might yield a non-integer result.

The numbers we create by dividing integers are called *rational* numbers. For example, 1.5 is rational because it is equal to $3 \div 2$.

The integer 3 is rational because $3 = 3 \div 1$ (it's also equal to $6 \div 2$, $12 \div 4$, and so on). All integers are rational numbers.

Whereas the integers "play nicely" with three of the basic operations, the rational numbers are fully compatible with all four. Given two rational numbers, their sum, difference, product, and quotient are all also rational numbers (with the usual caveat that division by 0 is forbidden).

The rational numbers are completely sufficient for everyday use. All of the quantities we measure—such

> Division by zero is hopeless; here we refer to calculations such as $7 \div 5$.

> The term *rational numbers* derives from the word *ratio* and is not a value judgment on their sanity.

as weights, volumes, distances, prices, temperatures, time, populations, radio frequencies—are accurately specified using rational numbers.

Since the rational numbers are sufficient for all practical work and are fully compatible with the basic operations of arithmetic, why do we need any other numbers?

But let's ask a more fundamental question: Are there any others?

The diagonal of a square

HOW FAR APART ARE THE OPPOSITE CORNERS OF A SQUARE? Later, in Chapter 14, we'll work through the solution to this problem. But for now, it is enough for us to simply report that the length of the diagonal of a 1×1 square is $\sqrt{2}$.

The number $\sqrt{2}$ has the property that if you multiply it by itself (that is, if you square it), then the result is 2. You can compute its value on most calculators, but let's see if we can determine its value with some simple calculations.

Here is a quick reminder of the meaning of square root. The square root of 9 is 3 because $3 \times 3 = 9$. We write this: $\sqrt{9} = 3$. In general, for a number N, the square root of N is a quantity that when multiplied by itself (squared) gives N. Algebraically, $a = \sqrt{N}$ means $a^2 = a \times a = N$.

To start, notice that if we square zero we get $0^2 = 0$ and if we square one we get $1^2 = 1$. These results are less than our target: 2. On the other hand if we square two we get $2^2 = 4$, if we square three we get $3^2 = 9$, and so forth. These results are all greater than our desired target.

Because 1^2 is too small but 2^2 is too large, the value of $\sqrt{2}$ must be bigger than 1 and less than 2. Let's look at the values between 1 and 2 in 0.1 increments as shown in the table.

We see that 1.4 is too small to be the square root of 2 and 1.5 is too large. So the value of $\sqrt{2}$ must lie between these.

Let's refine this once again. If we square the numbers between 1.4 and 1.5 in 0.01 steps we find that

$$1.41^2 = 1.9881 \quad \text{and} \quad 1.42^2 = 2.0164,$$

x	x^2
1.0	1.00
1.1	1.21
1.2	1.44
1.3	1.69
1.4	1.96
1.5	2.25
1.6	2.56
1.7	2.89
1.8	3.24
1.9	3.61
2.0	4.00

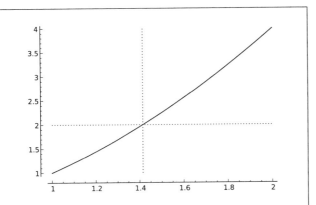

We can estimate the value of $\sqrt{2}$ by plotting the graph of the equation $y = x^2$ and checking where that curve crosses the line $y = 2$. As can be seen in this graph, that happens when x is just a bit larger than 1.4.

from which we may conclude that $1.41 < \sqrt{2} < 1.42$.

We can repeat this process, narrowing down the value of $\sqrt{2}$ to lie in tighter and tighter ranges.

At some point we either are satisfied (because we have achieved a fantastically accurate approximation for $\sqrt{2}$) or we are frustrated (because we have yet to figure out the value of $\sqrt{2}$ exactly).

But exactly what do we mean by *exactly*?

Beyond the rational

A REASONABLE NOTION OF BEING "EXACT" is the ability to specify a quantity as a rational number, that is, as the ratio of two integers. If we could express $\sqrt{2}$ as a fraction $\frac{a}{b}$ where a and b are integers, then we can rightfully say we have found its exact value.

Unfortunately, this is not possible, as we shall prove.

Theorem. $\sqrt{2}$ *is not a rational number.*

The proof that $\sqrt{2}$ is not rational uses the same "What if?" approach we used in Chapter 1 to show that there are infinitely many primes. We *suppose* that $\sqrt{2}$ is a rational number and see that this leads us to an impossible conclusion. The consequence: our supposition that the $\sqrt{2}$ is rational is untenable, and therefore $\sqrt{2}$ is not a rational number.

LET US PROCEED DIRECTLY WITH THIS APPROACH. Suppose that $\sqrt{2}$ were a rational number. That means that $\sqrt{2}$ is the ratio of two positive integers—let's call them a and b. We have

> We may assume that both a and b are *positive* integers because $\sqrt{2}$ is a positive quantity.

$$\frac{a}{b} = \sqrt{2}.$$

That a/b is the square root of 2 means that if we square this number, the result is 2. In symbols, we have

$$\left(\frac{a}{b}\right)^2 = 2. \tag{A}$$

Multiplying fractions is as easy as multiplying their numerators and denominators and so we have

$$\left(\frac{a}{b}\right)^2 = \frac{a}{b} \times \frac{a}{b} = \frac{a \times a}{b \times b} = \frac{a^2}{b^2}. \tag{B}$$

Taken together, equations (A) and (B) give

$$2 = \frac{a^2}{b^2},$$

which can be rewritten as

$$2b^2 = a^2. \tag{C}$$

Let's examine the two sides of equation (C). Since a is a positive integer, we can factor a into primes (see the Fundamental Theorem of Arithmetic on page 10); say

$$a = p_1 \times p_2 \times \cdots \times p_n$$

where p_1, p_2, and so on are primes.

Likewise, b is a positive integer and so it can be factored into primes like this:

$$b = q_1 \times q_2 \times \cdots \times q_m$$

where q_1, q_2, and so forth are primes.

The left-hand side of equation (C) can be rewritten as

$$2b^2 = 2 \times (q_1 \times q_2 \times \cdots \times q_m)^2$$
$$= 2 \times (q_1 \times q_1) \times (q_2 \times q_2) \times \cdots \times (q_m \times q_m).$$

Notice that this expresses the integer $2b^2$ as a product of an odd number of primes.

Similarly, the right-hand side of equation (C) can be rewritten as

$$a^2 = (p_1 \times p_2 \times \cdots \times p_n)^2$$
$$= (p_1 \times p_1) \times (p_2 \times p_2) \times \cdots \times (p_n \times p_n).$$

Notice that this gives a factorization of the integer a^2 into an even number of primes.

And now for the punch line: The two numbers $2b^2$ and a^2 are the same—that's what equation (C) says. We have shown that this integer (whatever its value may be) has one factorization into an odd number of primes and another factorization into an even number of primes. That's impossible because the Fundamental Theorem of Arithmetic says that there is only one way to factor a positive integer into primes.

We have reached an impossible situation. The root cause of this contradiction is the supposition that $\sqrt{2}$ is rational. Since that supposition leads us to an impossible conclusion, it must be false. Thus, $\sqrt{2}$ is *not* a rational number.

Since $\sqrt{2}$ is not a rational number we say that it is *irrational*. The rational numbers are quite adequate for specifying *physical* quantities we might consider, but they do not capture all *mathematical* quantities. The length of the diagonal of a 1×1 square is not a rational number.

Constructible numbers

STARTING WITH THE NUMBER 1 and then using the three operations addition, subtraction, and multiplication repeatedly, we can create any integer, and we create only integers. If we also include division as an operation, then we can create all possible rational numbers (and only the rational numbers).

What we have learned is that when we also include the square-root operation, we arrive at numbers that are not simple ratios of whole numbers. When we build nasty expressions such as

$$\frac{\sqrt{17}-1}{2+\sqrt{5}} \times \sqrt{1+\sqrt{7/3}}$$

we don't expect the result to be a rational number.

It's handy to give a name to the numbers we can build in this manner; let us say that a number is *constructible* if it can be calculated starting with the number 1 and applying the five operations $+$, $-$, $\times$, $\div$, and $\sqrt{}$ as often as we choose but with the standard caveats that we may not divide by zero and we may not take the square root of a negative quantity.

Which leads naturally to the question: Are all numbers constructible?

THE ANCIENT GREEKS SAW an intimate and beautiful connection between geometry and numbers. The link is established using two tools: the straight edge (a ruler without markings) and the compass. Given a unit length (that is, a line segment whose length is equal to one), what other lengths can we construct using a straight edge and compass?

It's not hard to describe how to add and subtract using the construction tools. Given line segments of lengths a and b, we use the straight edge to extend the first segment whose length is a. Then, opening the compass to match the length of the second seg-

> In this chapter we only consider square roots of nonnegative numbers. In Chapter 5 we explore the consequence of applying the square-root operation to negative numbers.

> Sometimes, of course, the result *may* be a rational number; for example,
>
> $$\left(\sqrt{7}-\sqrt{2}\right) \times \left(\sqrt{7}+\sqrt{2}\right)$$
>
> is equal to 5.

ment, we place the point of the compass at one end of the first segment and then use the pencil-end of the compass to mark a point distance b further down the line. The combined segment we have just created has length $a + b$. Subtraction is achieved by shortening, rather than extending, a segment.

The other operations are more complicated to perform using straight edge and compass but, in fact, with these tools we are able to multiply, divide, and extract square roots of lengths.

Indeed, the lengths we can form using straight edge and compass are precisely the (nonnegative) constructible numbers!

There was a time that the ancient Greeks believed that all numbers were rational, but their Pythagorean School worked out that this is not so.

Still, the Greeks clung to their aesthetically appealing belief in a link between geometry and numbers: the notion that all lengths—all (positive) numbers—could be constructed with straight edge and compass.

Belief in this equivalence is directly related to two of the three famous Greek construction problems. The best known of these is the *angle trisection problem*: Given an angle, use the construction tools to divide the angle into three equal parts.

Less famous, but on point for this discussion, are these conundrums:

- *Doubling the cube*. Given the length of the side of a cube, construct the length of the side of a cube that has twice the volume.

  Since a unit-length line segment is assumed to be given, this problem is tantamount to constructing a line segment whose length is $\sqrt[3]{2}$.

- *Squaring the circle*. Given a circle, construct a square with the same area as the circle.

  Again, since a unit length is given, one can con-

The angle *bisection* problem is easier. It is not difficult to use straight edge and compass to find the ray that divides an angle into two equal parts.

Despite the fact that mathematicians have known for over a century that angle trisection with straight edge and compass is impossible, enthusiasts continue to propose "solutions" to this problem. Some of their clever attempts are recounted here: Underwood Dudley, *A Budget of Trisections* (Springer, 1978).

struct a circle whose area is π (see Chapter 6). Squaring the circle is then equivalent to constructing a segment whose length is $\sqrt{\pi}$.

It would take two millennia for these problems to be solved. Neither $\sqrt[3]{2}$ nor $\sqrt{\pi}$ is constructible. Similarly, the solution to the angle trisection problem involves showing that a specific number (the cosine of 20°) is not constructible.

The existence of nonconstructible numbers breaks the hoped-for linkage between geometry and number that the Greeks sought to forge with straight edge and compass.

In 1837 Pierre Wantzel proved that these three numbers are not constructible. The implication is that none of the three classical construction problems are possible. Stated differently, Wantzel solved the angle trisection problem: he proved that there is no construction method for trisection.

Playing in tune

WHEN MUSICIANS PLAY INSTRUMENTS THAT ARE NOT IN TUNE with each other, the result is acoustic dissonance—the music sounds "off."

If two performers play the same note, the frequencies produced by their instruments should be the same; it is the discrepancy that bothers the listener. However, musicians often play different notes, and the music sounds best when the notes harmonize with each other. What makes for good harmony? What sounds good to our ears?

The Greeks considered this problem and discovered that notes whose frequencies are in small whole number ratios (e.g., 3 : 2) sound pleasing together. On this basis, a musical scale (attributed to Pythagoras) was invented. In figuring out frequencies for notes, there is one overarching constraint: Notes that are a full octave apart have frequencies in precisely a 2 : 1 ratio. To create harmonious sounds, the Greeks also desired that the frequency ratio between C and F and between C and G be expressed with small integers. In their solution, the ratio of the frequencies of adjacent notes was 9/8 for whole steps (e.g., from C to D) or 256/243 for half steps (e.g., from E to F).

Here is the entire Pythagorean scale:

$$C \xrightarrow{\frac{9}{8}} D \xrightarrow{\frac{9}{8}} E \xrightarrow{\frac{256}{243}} F \xrightarrow{\frac{9}{8}} G \xrightarrow{\frac{9}{8}} A \xrightarrow{\frac{9}{8}} B \xrightarrow{\frac{256}{243}} C.$$

From this we can calculate the relative frequency of, say, the notes C and F. To get F's frequency, we multiply C's frequency by

$$\frac{9}{8} \times \frac{9}{8} \times \frac{256}{243} \times \frac{9}{8} = \frac{4}{3}.$$

The 4 : 3 ratio of frequencies sounds great.

We can visualize the waveform when a C and an F are played together in this tuning system. The sound looks like this:

But when the F is tuned to a slightly higher pitch, the waveform looks like this:

The difference, which is noticeable to your eye, is also noticeable to your ear; you are seeing harmonic dissonance.

One weakness of the Pythagorean scale is that the ubiquitous C-major chord, C-E-G, is dissonant; the ratios of the frequencies are not simple.

Over the centuries, alternative tunings were developed. For example, in *just intonation*, the following ratios are used:

$$C \xrightarrow{\frac{9}{8}} D \xrightarrow{\frac{10}{9}} E \xrightarrow{\frac{16}{15}} F \xrightarrow{\frac{9}{8}} G \xrightarrow{\frac{10}{9}} A \xrightarrow{\frac{9}{8}} B \xrightarrow{\frac{16}{15}} C.$$

With this tuning, the C-E-G triad is a combination of frequencies in a lovely 4 : 5 : 6 ratio. But in this system, the whole step from C to D sounds different than the whole step for D to E.

The notation

$$C \xrightarrow{\frac{9}{8}} D$$

means that D's frequency is 9/8 times higher than that of C.

Both of these systems (Pythagorean and just intonation) have another serious problem. If a group of musicians finishes playing a piece in, say, C major and now wants to play a piece written in the key of F, their instruments need to be re-tuned. This is mildly inconvenient for the lute player, highly laborious for the harpsichordist, and essentially impossible for the woodwind players.

The solution is to create a tuning system that works equally well in all keys. This leads to two constraints: (a) notes an octave apart must have frequencies in a 2 : 1 ratio and (b) the frequency ratio for notes a half step apart is constant (e.g., the ratio of frequencies for C and $C^\#$ is the same as for $C^\#$ and D). There are twelve half steps in an octave:

$$C - C^\# - D - D^\# - E - F - F^\# - G - G^\# - A - A^\# - B - C.$$

If the frequency ratio between consecutive notes is some number r [constraint (b)] and traversing an octave doubles the frequency [constraint (a)], then we must have $r^{12} = 2$. This implies

$$r = \sqrt[12]{2} \approx 1.059463.$$

If we tune our instruments so that the ratio between adjacent notes is $\sqrt[12]{2}$ then no re-tuning is needed when switching from one key to another. This system of tuning is called *equal temperament*, and it is the system that is nearly universally used today.

UNFORTUNATELY, $\sqrt[12]{2}$ IS IRRATIONAL. That means the ratio of frequencies between notes is never a whole number (except for octaves). The frequencies for a C–G interval are not in a 3 : 2 ratio; rather, the ratio works out to be about 1.4983, which, admittedly, is pretty close to 1.5.

But how does it sound? At present, nearly all music is performed on instruments tuned to equally tempered scales, so the harmonies we hear are familiar to us. But let's see (literally) what we are missing.

The proof that $\sqrt[12]{2}$ is irrational is nearly identical to the proof that $\sqrt{2}$ is irrational. Try working it out.

Here are the waveforms for a C-major chord. In the first, the notes are played in precisely a 4 : 5 : 6 ratio of frequencies and in the second, the frequencies are as specified by equal temperament. The first waveform looks (and sounds!) much nicer than the second.

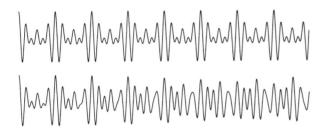

The advantage of equal temperament is that retuning instruments is not necessary between pieces. But there is one instrument capable of instant retuning: the human voice.

Unaccompanied singing groups, such as barber shop quartets, are not obliged to use equally tempered scales and can "bend" their notes so the frequencies are in small integer proportions. The result is the fantastic, resonant sound these groups produce.

5
i

Another square-root conundrum

IN CHAPTER 4 WE PONDERED THE "EXACT" VALUE of $\sqrt{2}$ and were led to the conclusion that we cannot express $\sqrt{2}$ as a ratio of two integers—$\sqrt{2}$ is irrational. We can, however, give excellent decimal approximations for $\sqrt{2}$.

Although $\sqrt{2}$ is not a rational number, we didn't question whether or not such a number x so that $x^2 = 2$ even exists. In fact, $\sqrt{2}$ is a legitimate number that lies somewhere between 1.41 and 1.42. It is an example of a *real* number. A real number can be expressed like this:

$$\pm XXXXX.XXXXXXXXXX\ldots$$

where the Xs are digits. The number begins with either a $+$ or $-$ sign (although the $+$ is usually omitted), finitely many digits before the decimal point, and then infinitely many digits after. For example, the number $1\frac{2}{3}$ can be written

$$1.666666666\ldots$$

Numbers such as $\frac{3}{4}$ whose decimal expansion is finite (0.75) can be written this way; we simply tack on an endless stream of 0s to the end: $0.7500000\ldots$.

The collection of real numbers is denoted $\mathbb{R}$.

See Chapter 3 for a more extensive discussion of repeating decimals.

Thus $\sqrt{2}$ is a real number; it just doesn't happen to be a rational number. Stated differently, there's a real number x such that $x^2 = 2$. Likewise, there's a real number that satisfies the equation $x^2 = 3$, namely $\sqrt{3} = 1.73205\ldots$ And so forth. Maybe.

DOES EVERY EQUATION OF THE FORM $x^2 = a$ HAVE A SOLUTION? If a is a positive real number (or zero) then $\sqrt{a}$ is a solution and the decimal expression can be calculated to as many digits as we desire. One way to visualize this is to plot the graph of $y = x^2 - a$ and see where the curve (which will be a parabola for an equation of this form) crosses the x-axis. That will be a number such that $x^2 - a = 0$ or, equivalently, $x^2 = a$. The first figure shows the graphs of $y = x^2 - 3$ and $y = x^2 - 7$. The two parabolas cut the x-axis at $\pm\sqrt{3}$ and $\pm\sqrt{7}$, respectively.

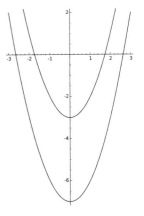

The problem changes considerably when we seek a number x such that $x^2 = -1$. Is there such a number? If we square a positive number, the result is again a positive number; for example, $5^2 = 5 \times 5 = 25 > 0$. Likewise, if we square a negative number, the result is also positive: $(-5)^2 = -5 \times -5 = 25 > 0$. And if we square 0 the result is zero. The situation looks hopeless.

Indeed, our despair is deepened if we plot the graph of the equation $y = x^2 + 1$ and search for where the parabola crosses the x-axis. Such a graph is shown in the second figure. We quickly see that this parabola lies entirely above—and never crosses— the x-axis.

We are tempted to give up hope and declare "there is no square root of negative one." But that attitude lacks imagination. True, there is no *real* number that satisfies $x^2 = -1$, but perhaps there's some other sort of number that does.

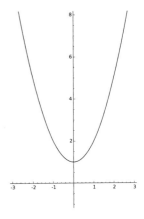

Imaginary numbers

THE SOLUTION TO THIS PROBLEM IS STARTLINGLY SIMPLE. There is no real number x for which $x^2 = -1$; so we simply make up a new number, and call it i, with the requisite property: $i^2 = -1$.

Immediately alarm bells go off. "Where did this number come from?" "You can't just make up numbers!" "This doesn't make sense!"

To ease our mind, we call i an *imaginary* number. Suddenly i is demoted to second-class citizenship, and we no longer expect to put i pencils in a coffee cup on our desk or worry that someone will tell us that the distance to some location is i miles.

We just agree to play the game. OK, we think. If I accept this number i, let's see what it can do. I know that $i \times i = -1$. What about $i + i$? Following the standard rules of arithmetic, the result is another imaginary number: $2i$. We might wonder, what happens if we square that number. Try it!

$$(2i)^2 = (2i) \times (2i) = 2 \times i \times 2 \times i = 2 \times 2 \times i \times i$$
$$= 4 \times (i \times i) = 4 \times -1 = -4.$$

In other words, $2i$ is a square root of -4.

We can square $\sqrt{2}i$; let's see what we get:

$$\left(\sqrt{2}i\right)^2 = \sqrt{2} \times i \times \sqrt{2} \times i = 2 \times i \times i = -2,$$

and so we see that $\sqrt{2}i$ is a square root of -2. Indeed, once we accept i into the family of numbers, we not only have a square root of -1, but we have square roots of all negative real numbers! Any number of the form bi, where b is a real number, is called an *imaginary* number.

If we add two imaginary numbers, such as $4i$ and $2i$, we get another imaginary number: $6i$. If we multiply two imaginary numbers, such as $3i$ and $-2i$, we get a real number:

$$3i \times -2i = 3 \times -2 \times i \times i = -6 \times -1 = 6.$$

So-called *real numbers* are no more "real" than imaginary numbers. We can't put -3 pencils in our coffee cup nor can we say that the distance to some location is *exactly* $\sqrt{2}$ miles. True, real numbers are useful for modeling certain physical phenomena such as temperature or land area. But imaginary numbers are useful for others, including quantum mechanics and electronics. All numbers are "imaginary" in the sense that they are the inventions of the mind.

Complex numbers

TO FULLY INCORPORATE IMAGINARY QUANTITIES into the family of numbers, we need to add, subtract, multiply, and divide real and imaginary numbers together. The framework for doing this is called the *complex number* system. This is an extension of the real numbers and includes all numbers of the form $a + bi$ where a and b are real, like this: $3 + 4i$.

The number i is a complex number because we can write it as $0 + 1i$. Likewise, any real number, such as -7, is a complex number because it can be written as $-7 + 0i$.

Addition of complex numbers is simple; we just collect like terms:

$$(3 + 2i) + (4 - 3i) = (3 + 4) + (2 - 3)i = 7 - i,$$

which can be written more pedantically as $7 + (-1)i$.

Subtraction is scarcely more involved:

$$(3 + 2i) - (4 - 3i) = (3 - 4) + (2 - (-3))i = -1 + 5i.$$

A moment's reflection reveals that the sum or difference of two complex numbers is also a complex number. We can write this algebraically like this:

$$(a + bi) + (c + di) = (a + c) + (b + d)i$$
$$(a + bi) - (c + di) = (a - c) + (b - d)i$$

where a, b, c, d are real numbers.

Multiplication of complex numbers is more challenging. Let's work through multiplying $3 + 2i$ and $4 - 3i$.

$$\begin{aligned}(3 + 2i) \times (4 - 3i) &= 3 \times (4 - 3i) + 2i \times (4 - 3i) &&\text{distributive property}\\ &= \left(3 \times 4 - 3 \times 3i\right) + \left(2i \times 4 - 2i \times 3i\right) &&\text{distributive again}\\ &= \left(12 - 9i\right) + \left(8i + 6\right)\\ &= 18 - i.\end{aligned}$$

Algebraically, if $a + bi$ and $c + di$ are two complex numbers, their product is given by this formula:

$$(a + bi) \times (c + di) = (ac - bd) + (ad + bc)i,$$

which implies that when we multiply two complex numbers, the result is also a complex number.

Division is the most complicated of the basic operations. Rather than diving headlong into an explanation of $(a + bi) \div (c + di)$ we first consider the *reciprocal* of a complex number. Recall that the reciprocal of a number x is another number y so that $xy = 1$. For example, $\frac{1}{2}$ is the reciprocal of 2.

What is the reciprocal of $1 + 2i$? We need a number $a + bi$ so that $(1 + 2i) \times (a + bi) = 1$. We show that $\frac{1}{5} - \frac{2}{5}i$ meets that requirement:

$$(1 + 2i) \times \left(\frac{1}{5} - \frac{2}{5}i\right) = 1 \times \left(\frac{1}{5} - \frac{2}{5}i\right) + 2i \times \left(\frac{1}{5} - \frac{2}{5}i\right)$$
$$= \left(\frac{1}{5} - \frac{2}{5}i\right) + \left(\frac{2}{5}i + \frac{4}{5}\right) = \frac{1}{5} + \frac{4}{5} = 1.$$

In general, we have the following formula for the reciprocal of $a + bi$:

$$\left(\frac{a}{a^2 + b^2}\right) - \left(\frac{b}{a^2 + b^2}i\right). \qquad (A)$$

To see that this is correct, one multiplies expression (A) by $a + bi$, works methodically through the algebra, and finds that the result is 1.

Notice that the denominators in (A) are both $a^2 + b^2$. If this happens to equal 0, then the formula is invalid since one cannot divide by zero. However, $a^2 + b^2 = 0$ only in case both a and b are 0. In other words, all complex numbers except $0 + 0i$ have reciprocals. And this is precisely what we would hope for: Just as 0 is the only real number that has no reciprocal, 0 is also the only complex number without a reciprocal. The reciprocal of any nonzero complex number is a complex number.

With reciprocals under control, we can finally consider division. Dividing a number X by Y is the same as multiplying X by the reciprocal of Y. It now follows that the quotient of two complex numbers (as long as the divisor is not 0) is again a complex number.

Here's the conclusion: The basic operations $+$, $-$, $\times$, and $\div$ all work perfectly with complex numbers. We can apply these operations to any pair of complex numbers (except division by zero) and the result is a complex number.

BUT NOW LET'S RETURN TO THE OPERATION that caused us trouble in the first place: square root. The real numbers were "deficient" in the sense that some numbers had square roots, but others did not. So we extended the reals by creating a new number $i = \sqrt{-1}$. We then applied arithmetic operations and the real number system grew to become the complex number system. But have we solved the square root problem? What about $\sqrt{i}$? Do we need to make up another new number, incorporate that, and create a monstrous "supercomplex" number system?

Thankfully the complex numbers are already rich enough to provide us with all the square roots we need! Let's see how we can find a square root of i without creating any new numbers.

We are looking for a complex number $a + bi$ with the property that $(a + bi)^2 = i$. Let's do a bit of algebra to figure this out. We start by expanding $(a + bi)^2 = (a + bi) \times (a + bi)$:

$$(a + bi)^2 = (a + bi) \times (a + bi) = (a^2 - b^2) + (2ab)i.$$

To make this expression equal to $i = 0 + 1i$ we need

$$a^2 - b^2 = 0 \quad \text{and} \quad 2ab = 1.$$

We find the solution to these equations as follows. The first equation, $a^2 = b^2$, implies that either $a = b$ or $a = -b$.

If $a = b$, then $2ab = 1$ can be rewritten $2a^2 = 1$. Dividing by 2 gives $a^2 = \frac{1}{2}$ and so

$$a = \frac{1}{\sqrt{2}} \quad \text{or} \quad a = -\frac{1}{\sqrt{2}}.$$

Since $a = b$, we find that both

$$\frac{1}{\sqrt{2}} + \frac{1}{\sqrt{2}}i \quad \text{and} \quad -\frac{1}{\sqrt{2}} - \frac{1}{\sqrt{2}}i$$

are square roots of i. You can verify this by squaring these and see that in both cases the result is i.

The other case, $a = -b$, leads to the same two solutions.

With some more work, one can show that all complex numbers have square roots, so there is no need to extend the complex numbers to accommodate the square root operation.

The Fundamental Theorem of Algebra

WHAT ABOUT CUBE ROOTS? A cube root of a number c is a number x such that $x^3 = c$. Does every complex number have a cube root within the complex number system, or do we need to invent new numbers?

Here's a challenge: Find the cube roots of i. One of them is $-i$ because $(-i) \times (-i) \times (-i) = -i^3 = i$. What are the other two? Answers appear on the next page.

The equation $x^3 = c$ can be rewritten $x^3 - c = 0$. The cube-root question can be extended more broadly. Does every *polynomial* equation have a complex number solution? For example: Is there a complex number x such that

$$3x^5 + (2-i)x^4 + (4+i)x^3 + x - 2i = 0?$$

A landmark result in the theory of complex numbers is that every polynomial equation has a solution! This result is known as the Fundamental Theorem of Algebra. The technical term used to describe this situation is that the complex numbers are *algebraically complete*.

Here is a complete statement of this important fact:

Theorem (Fundamental Theorem of Algebra). *Let d be a positive integer and let $c_0, c_1, c_2, \ldots, c_d$ be complex numbers with $c_d \neq 0$. Then there is a complex number z such that*

$$c_d z^d + c_{d-1} z^{d-1} + \cdots + c_2 z^2 + c_1 z + c_0 = 0.$$

The real numbers are, in a sense, deficient because there are some problems involving just the basic operations that cannot be solved (such as: find a number a such that $a \times a + 1 = 0$). The Fundamental Theorem of Algebra says, roughly, that in the domain of complex numbers, all such problems have solutions. *Finding* those solutions is a different matter!

Cube roots of i. In addition to $-i$, we have the following:

$$\frac{\sqrt{3}}{2} + \frac{1}{2}i \quad \text{and} \quad \frac{-\sqrt{3}}{2} + \frac{1}{2}i.$$

6

π

What is π?

THE NUMBER π HAS FASCINATED PEOPLE for generations. It has made its way into popular culture (both as the title of a movie and as the name of a cologne). Schoolchildren celebrate π day by competing to see who has memorized the most digits of π.

March 14 is celebrated as Pi Day because the date is written as 3/14 and π is approximately 3.14.

Pi is the sixteenth letter of the Greek alphabet. It is the ratio of a circle's circumference to its diameter. In other words, the circumference of a circle is always π times larger than its diameter; in symbols: $C = \pi d$ where C is the circumference of the circle and d is its diameter. This relation can also be expressed as $C = 2\pi r$ where r is the radius of the circle. Pi also is used to calculate the area of a circle via the formula $A = \pi r^2$ where r is the radius. Not surprisingly, π appears in the formulas for the surface area and volume of a sphere. If the radius of a sphere is r, its surface area is $4\pi r^2$ and its volume is $\frac{4}{3}\pi r^3$.

These formulas from geometry don't tell us the numerical value of π. A simple argument shows that π is larger than 3. Draw a circle with radius equal to 1 and inscribe in that circle a regular hexagon. Then carve up that hexagon into six equilateral triangles (as shown in the figure). Because the circle's radius

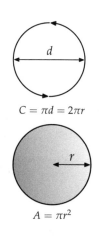

$C = \pi d = 2\pi r$

$A = \pi r^2$

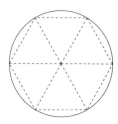

is 1, the lengths of the sides of all the triangles are also equal to 1. It follows that the perimeter of the hexagon is 6. The circumference of the circle (which equals 2π because the radius is 1) is a little bigger than the perimeter of the hexagon, and so we have that
$$2\pi > 6.$$
Dividing both sides by 2 gives us $\pi > 3$. We can see that the circle's circumference isn't too much bigger than the perimeter of the hexagon, so π is not too much bigger than 3.

On the other hand, we can circumscribe a regular hexagon around the outside of the circle, and carve that hexagon into six equilateral triangles. In this case, a bit of geometry shows that the side length of each equilateral triangle—and hence the side length of the hexagon—is $2/\sqrt{3}$. (This is a good exercise in the use of the Pythagorean Theorem—see Chapter 14. The solution is given on page 59.)

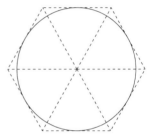

It follows that the perimeter of the surrounding hexagon is six times as large: $6 \times 2/\sqrt{3} = 12/\sqrt{3}$. Notice that the circumference of the circle (2π) is less than the perimeter of the hexagon ($12/\sqrt{3}$). Therefore
$$2\pi < \frac{12}{\sqrt{3}}$$
and dividing by 2 yields
$$\pi < \frac{6}{\sqrt{3}} = 3.464\ldots$$

Instead of inscribing and circumscribing a radius-one circle with a regular hexagon, we may choose to use a regular n-gon where n is much larger than 6. For example, if we take $n = 100$ and then calculate (using some trigonometry) the perimeters of inscribed and circumscribed regular 100-gons, we get these estimates:
$$3.1410759 < \pi < 3.1426266.$$

In the limit, as the number of sides of the inscribed and circumscribed regular polygons grows toward infinity, we find that the approximate value of π is

$$\pi = 3.14159265358979323846264338327950 2884\ldots$$

WHAT IS THE "EXACT" VALUE OF π? In Chapter 4, we explained that $\sqrt{2}$ does not have an "exact" value. That is, $\sqrt{2}$ is irrational—it cannot be expressed as the ratio of two integers. Likewise, π is irrational. Schoolchildren are sometimes taught that π "equals" 22/7, but this is only an approximation.

There are no simple expressions for π, but the following formulas nearly fit the bill:

$$\pi = 2 \times \left[\frac{2}{1} \times \frac{2}{3} \times \frac{4}{3} \times \frac{4}{5} \times \frac{6}{5} \times \frac{6}{7} \times \frac{8}{7} \times \frac{8}{9} \times \cdots \right] \quad (A)$$

$$\pi = 4 \times \left[\frac{1}{1} - \frac{1}{3} + \frac{1}{5} - \frac{1}{7} + \frac{1}{9} - \cdots \right]. \quad (B)$$

The fraction 22/7 evaluates to 3.142857. A better approximation for π is $355/113 = 3.14159292\ldots$

In both cases, one needs to take the calculation "to infinity," which, of course, is not feasible. Instead, one can stop the calculation after finitely many steps to acquire an approximation to π.

Neither formula (A) nor (B) gives a practical method for calculating π. For example, if we evaluate formula (A) through to the terms $\frac{200}{199} \times \frac{200}{201}$ we get $\pi \approx 3.134$. Likewise, if we calculate formula (B), stopping at the the terms $+\frac{1}{197} - \frac{1}{199}$, we get $\pi \approx 3.13159$.

More sophisticated techniques can be used to calculate approximations for π quickly and accurately. For science and engineering, an approximation of π that's accurate to, say, 30 digits is adequate for virtually any application. Just for the sheer joy and challenge, mathematicians and computer scientists have calculated π to more than one trillion decimal places.

Transcendence*

BOTH π AND $\sqrt{2}$ ARE IRRATIONAL but we can make a more profound assertion about π: it's *transcendental*. Let's see what that means.

Rational numbers are ratios of integers; examples include $\frac{5}{2}$, $\frac{-2}{3}$, and $\frac{7}{1}$. Here is another way to describe rational numbers: They are solutions to equations of the form

$$ax + b = 0$$

where a and b are integers. For example, $\frac{5}{2}$ is a solution to the equation $2x - 5 = 0$.

The number $\sqrt{2}$ is not rational (see Chapter 4) and therefore it is not the solution to a *linear* equation of the form $ax + b = 0$ where a and b are integers. However, it is the solution to a *quadratic* equation

$$ax^2 + bx + c = 0$$

where a, b, and c are integers. Specifically, it is a solution to this quadratic equation: $x^2 - 2 = 0$.

What about π? Since it's irrational, it is not the solution to a linear equation $ax + b = 0$ (where a and b are integers). Is it the solution to an integer-coefficient quadratic equation $ax^2 + bx + c = 0$? Again, the answer is no. Perhaps we have to appeal to a high power?

Is π the solution to an integer-coefficient *cubic* equation of the form $ax^3 + bx^2 + cx + d = 0$? Again, the answer is no. Quartic? Qunitic? Anythingic???

It turns out that π is *transcendental*. This means that π is not the solution to an integer-coefficient polynomial equation of any degree. In symbols, there is no polynomial equation of the form

$$a_n x^n + a_{n-1} x^{n-1} + \cdots + a_2 x^2 + a_1 x + a_0 = 0$$

(where all the coefficients a_k are integers) to which π is a solution.

> The rational number p/q is the solution to the equation $qx - p = 0$. Conversely, the solution to the equation $ax + b = 0$ (where a and b are integers) is the rational number $-b/a$.

Relatively prime

THE NUMBER π CROPS UP "BY SURPRISE" in parts of mathematics that have nothing to do with circles or even geometry. For example, π mysteriously appears in Stirling's formula for approximating factorials (see Chapter 10, page 106). Here we present another such instance that involves a basic property of integers: *relative primality*.

We say that two positive integers a and b are *relatively prime* provided that their only common divisor is 1.

For example, consider the numbers 15 and 28. Their divisors are as follows:

divisors of 15 → 1, 3, 5, and 15
divisors of 28 → 1, 2, 4, 7, 14, and 28.

Notice that the only divisor they have in common is 1; hence we say that 15 and 28 are relatively prime.

On the other hand, 21 and 35 are *not* relatively prime because both numbers are divisible by 7.

SUPPOSE WE ROLL TWO DICE. What is the probability that the two numbers that show are relatively prime?

Each die shows one of the values 1, 2, 3, 4, 5, or 6—all with equal probability. And whatever the first die shows, the second die also has six equally likely outcomes. All together, there are 36 possible results:

(1,1) (1,2) (1,3) (1,4) (1,5) (1,6)
(2,1) (2,2) (2,3) (2,4) (2,5) (2,6)
(3,1) (3,2) (3,3) (3,4) (3,5) (3,6)
(4,1) (4,2) (4,3) (4,4) (4,5) (4,6)
(5,1) (5,2) (5,3) (5,4) (5,5) (5,6)
(6,1) (6,2) (6,3) (6,4) (6,5) (6,6)

All of these 36 outcomes are equally likely. From this we can calculate, for example, the probability that the sum of the numbers on the dice is 7. There are six outcomes that give 7 as the sum: (1,6), (2,5), (3,4),

(4, 3), (5, 2), and (6, 1). Hence the probability the two numbers add to 7 is $6/36 = 1/6$.

We're asking a similar question: What is the probability that the two numbers are relatively prime? The following chart is useful for finding the answer. We place a star * in a given row and column of the chart if those two numbers are relatively prime. For example, there's a * in row 2 column 5 because 2 and 5 are relatively prime. However, there is no * in row 4/column 6 because 4 and 6 are not relatively prime.

	1	2	3	4	5	6
1	*	*	*	*	*	*
2	*		*		*	
3	*	*		*	*	
4	*		*		*	
5	*	*	*	*		*
6	*				*	

A bit of counting reveals 23 *s, so the probability that the two numbers are relatively prime is 23/36 which equals $0.638888\ldots$.

Let's repeat this exercise with twenty-sided dice. Two such dice are tossed and we ask: What is the probability that the numbers rolled are relatively prime? The solution is to make a much bigger chart! This new chart (shown on the next page) has 20 rows and 20 columns for a total of 400 cells. There's a star in row a/column b exactly when a and b are relatively prime. Painstaking examination of the chart reveals 255 stars, and so the probability that the two dice yield relatively prime values is 255/400, which equals 0.6375.

In general, we may ask, if we pick two numbers between 1 and N at random, what is the probability they are relatively prime? This can be calculated by computer by considering every pair of numbers between 1 and N: (1, 1), (1, 2), (1, 3), and so forth

Twenty-sided dice can be purchased in hobby shops. They are made in the shape of an icosahedron; see Chapter 16.

There is an efficient method to determine if two integers are relatively prime that we present in Chapter 12.

	1	2	3	4	5	6	7	8	9	10	11	12	13	14	15	16	17	18	19	20
1	*	*	*	*	*	*	*	*	*	*	*	*	*	*	*	*	*	*	*	*
2	*		*		*		*		*		*		*		*		*		*	
3	*	*		*	*		*	*		*	*		*	*		*	*		*	*
4	*		*		*		*		*		*		*		*		*		*	
5	*	*	*	*		*	*	*	*		*	*	*	*		*	*	*	*	
6	*				*		*				*		*				*		*	
7	*	*	*	*	*	*		*	*	*	*	*	*		*	*	*	*	*	*
8	*		*		*		*		*		*		*		*		*		*	
9	*	*		*	*		*	*		*	*		*	*		*	*		*	*
10	*		*				*		*		*		*				*		*	
11	*	*	*	*	*	*	*	*	*	*		*	*	*	*	*	*	*	*	*
12	*			*		*					*		*				*		*	
13	*	*	*	*	*	*	*	*	*	*	*	*		*	*	*	*	*	*	*
14	*		*				*				*		*		*		*		*	
15	*	*		*			*	*			*		*			*	*		*	
16	*		*		*		*				*		*		*		*		*	
17	*	*	*	*	*	*	*	*	*	*	*	*	*	*	*	*		*	*	*
18	*				*		*				*		*				*		*	
19	*	*	*	*	*	*	*	*	*	*	*	*	*	*	*	*	*	*		*
20	*		*				*		*		*		*				*		*	

until (N,N), and counting how many pairs are relatively prime. We finish by dividing that total by N^2 (the number of possible pairs). When we do this for various values of N we achieve the following results:

N	Probability
10	0.63
100	0.6087
1,000	0.608383
10,000	0.60794971
100,000	0.6079301507
1,000,000	0.607927104783

We see that as N tends toward infinity, the probability that two integers between 1 and N are relatively prime approaches a value around 0.6079. What is this limiting value? Remarkably, it is

$$\frac{6}{\pi^2} = 0.607927101854\ldots$$

Pi: It's not just about circles!

Solution to the problem on page 53.

A hexagon is circumscribed around a circle of radius 1. The hexagon is partitioned into six equilateral triangles. Show that the length of the sides of the triangles—and hence the hexagon—is $2/\sqrt{3}$.

Divide one of the equilateral triangles into two right triangles by drawing a radius from the center of the circle to the midpoint of one side of the hexagon as shown in the diagram. The midpoint divides the side of the triangle into two segments of length x (so the length of a side of the hexagon is $2x$). Since the six triangles are equilateral, all three sides of these triangles have length $2x$.

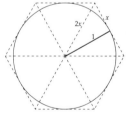

The legs of the right triangle have lengths 1 and x, and the hypotenuse has length $2x$. By the Pythagorean Theorem (page 155), we have

$$1^2 + x^2 = (2x)^2$$
$$1 + x^2 = 4x^2$$
$$1 = 3x^2$$
$$\frac{1}{3} = x^2$$

and so $x = 1/\sqrt{3}$. Since the perimeter of the hexagon is $12x$, we conclude that the perimeter is $12/\sqrt{3}$.

7
e

Leonhard Euler

IS THERE A GREATER HONOR FOR A MATHEMATI-
CIAN than to have a number named for you? Leonhard Euler (pronounced "oiler") was an eighteenth-century Swiss mathematician, and in this chapter we introduce the remarkable Euler number, e.

Euler did not name this number after himself, but he did choose the letter e as its symbol; historians doubt the selection was self-aggrandizing. Euler was a modest man of prodigious talent.

EULER'S NUMBER e CAN BE DEFINED in various ways but my preference is to express it as follows:

$$e = \frac{1}{0!} + \frac{1}{1!} + \frac{1}{2!} + \frac{1}{3!} + \frac{1}{4!} + \cdots \quad (A)$$

where the dots indicate that the sum goes on forever. Although there are infinitely many terms in the expression, its value is finite and it takes relatively few terms to arrive at a highly accurate approximation for e. For example, if we calculate this sum just to the $1/10!$ term, we find

$$\frac{1}{0!} + \frac{1}{1!} + \frac{1}{2!} + \cdots + \frac{1}{10!} = \frac{9864101}{3628800}$$
$$= 2.71828180114638\ldots,$$

which is very close to the actual value $2.718281828459045\ldots$.

The denominators in (A) are *factorials*. For a positive integer n, the value of $n!$ is

$$n \times (n-1) \times (n-2) \times \cdots \times 2 \times 1.$$

For example, $4! = 4 \times 3 \times 2 \times 1 = 24$. In addition, $0!$ is defined to be 1; this (and much more about the factorial function) is discussed in Chapter 10.

Unfortunately, there isn't a simple way to express e. In particular, e is not a rational number—see Chapter 4. Like π, it's a transcendental number; see page 55.

Euler's number is ubiquitous in mathematics. In this chapter we present three disparate problems each of whose solutions involves Euler's remarkable number, e.

An "interesting" number

A BANK OFFERS A TEN-YEAR CERTIFICATE OF DEPOSIT that, after the certificate matures, returns double the money invested. That is, if you deposit $1000 today, at the end of the term, the bank gives you $2000. That's 100% growth of your investment. It would be reasonable for the bank to advertise this CD as paying 10% interest per year (since it pays 100% interest after ten years).

The bank can sweeten the deal by paying out the interest before the end of the decade and reinvesting the proceeds back into the CD. Let's see what happens if the bank pays 10% interest at the end of each year and then reinvests that interest.

Starting with $1000: At the end of the first year, the CD has earned $100 in interest and so the value of the CD is now $1100. This larger sum now accrues interest during the second year. So instead of adding $100 at the end of year two, the bank adds 10% of $1,100 (that is, $110) to the account. So the value of the account is now $1210. At the end of year three, the bank adds 10% of that sum. Let's see exactly how the interest is posted and how the balance grows:

Year	Initial balance	Interest added	End-of-year balance
1	1000.00	100.00	1100.00
2	1100.00	110.00	1210.00
3	1210.00	121.00	1331.00
4	1331.00	133.10	1464.10
5	1464.10	146.41	1610.51
6	1610.51	161.05	1771.56
7	1771.56	177.16	1948.72
8	1948.72	194.87	2143.59
9	2143.59	214.36	2357.95
10	2357.95	235.80	2593.75

Each year the balance grows by 10%. In symbols, if the initial amount in a year is A, then the end-of-year balance is $1.1 \times A$. So two years' growth can be expressed as $1.1 \times 1.1 \times A$. Reasoning this way, the amount of money paid at the end of the term is

$$\underbrace{1.1 \times 1.1 \times \cdots \times 1.1}_{\text{ten terms}} \times 1000 = 1.1^{10} \times 1000 = 2593.74,$$

which agress with our earlier calculation. By compounding interest every year, our money grows not by a factor of two, but by a more generous factor of (about) 2.6.

There is a slight discrepancy between the value we calculate here and the previously computed value because each entry in the chart is rounded to the nearest cent.

WHAT HAPPENS IF THE BANK COMPOUNDS interest quarterly instead of annually? Since the interest rate on the certificate is 10% per year, the quarterly payment should be 2.5% of the balance. So in the first quarter, our initial $1000 investment grows by 2.5%, that is, 0.025×1000, which is $25. At the end of the first quarter, we have $1025. This is equivalent to multiplying our balance by 1.025.

In the second quarter, our $1025 account grows another 2.5%, adding another $25.63 to a new value of $1050.63, or, equivalently, $1025 \times 1.025 = 1050.63$ (to the nearest penny).

After N quarters, our initial $1000 is worth

$$\underbrace{1.025 \times 1.025 \times \cdots \times 1.025}_{N \text{ terms}} \times 1000 = 1.025^N \times 1000.$$

To calculate the value of our account after ten years, we take $N = 40$ in this formula (because in ten years there are 40 quarters) to find that our CD is worth $2685.06 at maturity.

Simple interest doubled our money. Annual compounding increased its value by a factor of 2.59 and quarterly compounding by a factor of 2.69. What happens if the bank compounds our interest monthly? Or even better, daily!?

For monthly compounding, the bank increases the value of our account by $10\% \div 12$ each month. Algebraically, if the balance at the beginning of the month is A, then at the end of the month the value of

In the first month, the bank will pay $10\% \div 12 = 0.8333\%$ of $1000, which is about $8.33. So the new balance is $1.00833 \times 1000 = 1008.33$.

the account is
$$\left(1 + \frac{0.1}{12}\right) \times A.$$

After N months the value of the CD is
$$\left(1 + \frac{0.1}{12}\right)^N \times 1000,$$

and if we substitute $N = 120$ (because there are 120 months in ten years) the value at maturity works out to be $2707.04.

The number of days in a year varies (because of leap years); to simplify the calculation for daily compounding, we assume every year has 365 days. If our bank compounds our interest daily, the value of our account should increase by 10% ÷ 365 each day. After N days, the CD's balance is
$$\left(1 + \frac{0.1}{365}\right)^N \times 1000.$$

Taking $N = 3650$ yields a balance at maturity of $2717.91.

But why wait until the end of the day to claim our interest? What if the bank paid interest every hour? ...every minute? ...every second!?

There are 31,556,926 seconds in a year, and so the compounded interest after ten years is

This accounts for the true length of a year being a bit different from 365 days.

$$\left(1 + \frac{0.1}{31556926}\right)^{10 \times 31556926} \times 1000$$

which evaluates to $2718.28.

Let's summarize our calculations in a single chart:

Compounding	Payout
Decade	$2000.00
Year	$2593.74
Quarter	$2685.06
Month	$2707.04
Day	$2717.91
Second	$2718.28

Why stop with seconds? The bank may choose to compound every tenth of a second, or every millisecond, or nanosecond. But compounding at fractions of a second doesn't actually change the bottom line. The payout stays as $2718.28 (but the small fractions of a penny that we ignore do change slightly).

In the limit, we reach the concept of *continuous* compounding. If the bank paid fractions of a cent, the exact payoff would be $1000 \times e$ dollars!

CONTINUOUS COMPOUNDING IS AN EXAMPLE of *exponential growth*. Let A be the initial amount of a substance (money, microbes, etc.). The substance grows at a rate r over a period of time t. If the newly produced substance also starts growing the instant it is produced (also at the rate r), then at the end of the time period, the total amount of the substance is

$$Ae^{rt}$$

where e is Euler's number. In our example, $A = 1000$ (initial deposit), $r = 0.1$ (the interest rate), and $t = 10$ (the decade), and so the closing balance is $1000 \times e$.

Likewise, substances can decay exponentially. If we take A to stand for the initial amount of the substance, r to be the rate of decay, and t the time period, the amount of substance at the end is Ae^{-rt}.

A good example of exponential decay arises in carbon-14 dating. The formula is used to calculate the age of a fossil by measuring the relative amount of carbon 14 present.

The deranged hat check clerk

IN A BYGONE ERA, AVID THEATER PATRONS would wear hats *en route* to a performance and would check their hats upon arrival at the theater. After the show, they would reclaim their hats.

One time, the hat check clerk—perhaps being a bit inebriated or perhaps being innately deranged—returns the hats at random to the people as they leave. The question is: *What is the probability that no one gets his or her own hat?*

Let's make sure the question is precise. There are N patrons and they wait in a line to receive their hats. The deranged clerk returns the hats in some order that is random. Since there are N patrons, there are $N!$ different orders in which the hats might be returned. We assume each of these orderings is equally likely. That's our interpretation of "at random."

See Chapter 10 where we discuss factorial notation and its relation to the problem of counting orderings.

For example, let us consider the case $N = 4$. The following chart shows all the ways in which the hats can be returned, and we flag with an arrow those distributions in which no one gets his or her own hat.

Person:	A	B	C	D	Person:	A	B	C	D
	A	B	C	D			C	A	B	D
	A	B	D	C	→	C	A	D	B
	A	C	B	D			C	B	A	D
	A	C	D	B			C	B	D	A
	A	D	B	C	→	C	D	A	B
	A	D	C	B	→	C	D	B	A
	B	A	C	D	→	D	A	B	C
→	B	A	D	C			D	A	C	B
	B	C	A	D			D	B	A	C
→	B	C	D	A			D	B	C	A
→	B	D	A	C	→	D	C	A	B
	B	D	C	A	→	D	C	B	A

Of the 24 distributions, there are 9 in which no one gets the proper hat. So the probability, when $N = 4$ is 9/24 (or 0.375 in decimal).

For $N = 5$ there are $5! = 120$ different ways to return hats. Among these, there are 44 in which no one gets his or her own. So the probability that no one gets the correct hat is 44/120 (or 0.3666... in decimal). The chart on the right shows the probabilities for larger values of N. It appears that the probability stops changing once N is 12, but, in fact, it does change slightly in decimal places beyond what we

N	Probability
6	0.3680556
7	0.3678571
8	0.3678819
9	0.3678792
10	0.3678795
11	0.3678795
12	0.3678794
20	0.3678794
100	0.3678794

show.

With some advanced analysis, we can derive a precise formula for the probability that when N theater patrons get their hats at random, the probability no one gets his or her own hat is precisely

$$\frac{1}{0!} - \frac{1}{1!} + \frac{1}{2!} - \frac{1}{3!} + \cdots \pm \frac{1}{N!}.$$

For example, when $N = 4$ this is

$$\frac{1}{1} - \frac{1}{1} + \frac{1}{2} - \frac{1}{6} + \frac{1}{24} = \frac{24 - 24 + 12 - 4 + 1}{24} = \frac{9}{24},$$

a result that agrees with our earlier analysis.

In the limit, as N grows to infinity, the probability no one gets the proper hat is

$$\frac{1}{0!} - \frac{1}{1!} + \frac{1}{2!} - \frac{1}{3!} + \frac{1}{4!} - \cdots \qquad (B)$$

where the terms go on without end. Notice the similarity of this expression to the formula for e given in equation (A). The sum in (B) evaluates to $1/e$, the reciprocal of Euler's number!

For $N = 10$, the value of (B) is precisely $16687/45360$, which equals $0.367879188712522\ldots$, which is quite near to the value $1/e = 0.367879441171442\ldots$.

The average gap between primes

IN CHAPTER 1 WE SHOWED THAT THERE are infinitely many primes. We noted that for small positive integers, primes appear quite often, but as we move to larger values, the primes "thin out." We can state with some precision just how "thin" the primes become by considering the *average gap between consecutive primes*.

Let's think about the primes between 1 and 20. They are

2, 3, 5, 7, 11, 13, 17, and 19.

For those familiar with logarithms: A way to measure how primes become sparser as we consider larger integers is to count the number of primes between 1 and a large integer N. A central result in number theory shows that the number of primes between 1 and N is nearly $N/\ln N$ where $\ln N$ is the base-e logarithm of N. This result is known as the Prime Number Theorem.

The gaps (differences) between these primes are

$$1, 2, 2, 4, 2, 4, \text{ and } 2$$

so the average gap is

$$\frac{1+2+2+4+2+4+2}{7} = \frac{17}{7} \approx 2.43.$$

Let's consider the same question for the primes between 1 and 1000. There are 168 primes starting with 2, 3, 5, and then ending 983, 991, and 997. The average gap between consecutive primes in this range is

$$\frac{(3-2)+(5-3)+(7-5)+\cdots+(991-983)+(997-991)}{167}.$$

The denominator is 167 because there are 168 primes, and therefore 167 gaps. The numerator can be calculated very easily. Notice that the $+3$ in the first term $(3-2)$ is cancelled by the -3 in the second term $(5-3)$. Likewise, the $+5$ in the second term is cancelled by the -5 in the third term, $(7-5)$. Indeed each "plus" value in a parenthesized term is cancelled by the "minus" value in the next term. After we make all these cancellations, the only terms left standing are the -2 from the first and the $+997$ from the last. Therefore the average gap between primes in the range 1 to 1000 is

$$\frac{997-2}{167} \approx 5.96,$$

which is more than twice the average gap between the primes up to 20.

FOR A POSITIVE INTEGER N, LET agap(N) STAND FOR the average gap between consecutive primes in the range 1 to N. For example, our earlier calculations can be written:

$$\text{agap}(20) = \frac{17}{7} \approx 2.43 \quad \text{and}$$
$$\text{agap}(1000) = \frac{995}{167} \approx 5.96.$$

The numerator in this expression is an example of what mathematicians call a *telescoping sum*. Imagine a hand-held spyglass composed of nested tubes. The telescope collapses for storage by sliding the sections into each other.

Similarly, each term in the numerator "collapses" into the next. So all the terms from $(3-2)$ up to $(997-991)$ fold up leaving the simple expression $(997-2)$.

Here's a chart of values of agap(*N*) for *N* equal to one hundred, one thousand, ten thousand, and so on, up to one billion:

N	agap(N)
10^2	3.958
10^3	5.958
10^4	8.120
10^5	10.425
10^6	12.739
10^7	15.047
10^8	17.357
10^9	19.667

Notice that each time *N* increases by a factor of 10, the value of agap(*N*) increases by about 2.3.

A way to illustrate this relationship is to plot, for a wide range of *N*, the value of agap(*N*). In the following graph, we plot for each value of *N* (on the horizontal axis) the corresponding value of agap(*N*) vertically with a small star. The vertical axis has standard spacing, but the horizontal axis is scaled so that each subsequent tick mark is 10 times more than the previous. This is known as a *logarithmic scale*.

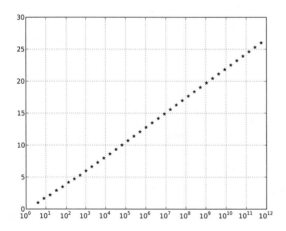

Notice that the stars lie (almost) exactly on a straight line. If you look carefully, the lower left portion is slightly curved upward.

If the stars in the graph were exactly on a line, then we would have the following equation in which we encounter Euler's number:

$$N = e^{a+1} \qquad (C)$$

where a is agap(N). For example, when $N = 10^{12}$, we have agap(N) = 26.59, which agrees closely with the value $a = 26.63$ needed to make equation (C) hold.

A miraculous equation

THIS CHAPTER AND THE PREVIOUS TWO have been devoted to three marvelous numbers: π, i, and e. Fantastically, these three numbers come together in a single equation (due to Euler):

$$e^{i\pi} + 1 = 0.$$

After being struck by this equation's awesome beauty and simplicity, a moment of mathematical anxiety takes hold: What does it mean to raise e to an imaginary exponent!?

We know what it means to raise e to a positive integer power. For example, $e^3 = e \times e \times e$. Negative powers are simply powers of the reciprocal: $e^{-2} = \frac{1}{e} \times \frac{1}{e}$. Fractional powers are explained in terms of square roots, cube roots, and so forth: $e^{1/2} = \sqrt{e}$. So even something as "dreadful" as $e^{-2/3}$ can be worked out:

$$e^{-2/3} = \frac{1}{\sqrt[3]{e}} \times \frac{1}{\sqrt[3]{e}}.$$

But $e^{i\pi}$ does not seem to fit into this framework at all. We need another approach.

Recall the e is defined by equation (A) to be

$$e = \frac{1}{0!} + \frac{1}{1!} + \frac{1}{2!} + \frac{1}{3!} + \frac{1}{4!} + \cdots.$$

It turns out that for any number x, the value of e^x is

$$e^x = \frac{1}{0!} + \frac{x}{1!} + \frac{x^2}{2!} + \frac{x^3}{3!} + \frac{x^4}{4!} + \cdots. \qquad (D)$$

We omit many steps in this derivation of Euler's formula; our intent is to explain the meaning of raising e to an imaginary power and to give a general outline of the proof of Euler's equation. To fill in the details requires some trigonometry and calculus that we cannot cover here.

For example, if we take $x = -1$, then equation (D) reduces to equation (B).

To evaluate $e^{i\pi}$ we substitute $i\pi$ for x in (D) to get

$$e^{i\pi} = \frac{1}{0!} + \frac{i\pi}{1!} + \frac{(i\pi)^2}{2!} + \frac{(i\pi)^3}{3!} + \frac{(i\pi)^4}{4!} + \cdots$$

Let's look closely at the numerators in this expression:

$$(i\pi)^2 = (i\pi) \times (i\pi) = i^2 \times \pi^2 = -\pi^2$$
$$(i\pi)^3 = i \times i \times i \times \pi^3 = -1 \times i \times \pi^3 = -i\pi^3$$
$$(i\pi)^4 = i^4 \times \pi^4 = \pi^4$$
$$(i\pi)^5 = -i\pi^5$$
$$(i\pi)^6 = -\pi^6$$
$$(i\pi)^7 = -i\pi^7$$
$$(i\pi)^8 = \pi^8$$
$$\vdots$$

Notice that the terms alternate between being real and imaginary. Collecting the real terms and the imaginary terms separately gives

$$e^{i\pi} = \left(\frac{1}{0!} - \frac{\pi^2}{2!} + \frac{\pi^4}{4!} - \frac{\pi^6}{6!} + \cdots \right) + \left(\frac{\pi}{1!} - \frac{\pi^3}{3!} + \frac{\pi^5}{5!} - \frac{\pi^7}{7!} + \cdots \right) i.$$

It turns out that the first parenthesized term is precisely $\cos \pi$ (which equals -1) and the second parenthesized term is $\sin \pi$ (which equals 0). It therefore follows that

$$e^{i\pi} = \cos \pi + i \sin \pi = -1 + 0i = -1$$

and Euler's fabulous formula follows forthwith.

8
∞

"To infinity ... and beyond!" is the brave battle cry of Buzz Lightyear, the fearless space ranger in Pixar's *Toy Story* movies. Buzz's catchphrase is funny because it's nonsensical: How can anything be "beyond" infinity? What could possibly be bigger than something infinite!? The apparent absurdity of the question likely prevented mathematicians from even uttering it, much less answering it. Yet this is precisely the question addressed by Georg Cantor in the late 1800s. Our overwhelming intuition is to answer "no": there is nothing "beyond infinity."

Our intuition, in this case, is wrong.

> Cantor's work was harshly criticized in its day on both mathematical and philosophical/religious grounds. However, it was not too long before his work became widely accepted and hailed as groundbreaking.

Sets

In mathematics, every concept is defined in terms of simpler concepts. In a rigorous, methodical development of mathematics, complex numbers are defined in terms of real numbers, real numbers in terms of rational numbers, rational numbers in terms of integers, and so forth. The tower of mathematics rests on one foundational concept: the set.

A *set* is simply a collection of things. For example, $\{1, 2, 5\}$ is a set that contains three numbers. It is the same as the set $\{2, 5, 1\}$ because the order in which we list elements is irrelevant. Furthermore, an

> The standard notation for a set is a list of its members enclosed in curly braces $\{\dots\}$.

object either is or is not a member of a set; it cannot be in a set more than once. Hence $\{1,1,2,5\}$ is the same as the set $\{1,2,5\}$; the extra appearance of 1 is superfluous.

Mathematicians use the symbol $\in$ to indicate membership in a set. For example, the notation $2 \in \{1,2,5\}$ means: "The number 2 is an element of the set $\{1,2,5\}$." A slashed $\in$ means that the object is not an element of the set; for example, $3 \notin \{1,2,5\}$.

For a set A, we write $|A|$ to stand for the number of elements in the set A. For example, $|\{1,2,5\}| = 3$. We call $|A|$ the size or *cardinality* of the set A.

A set such as $\{1,2,5\}$ has finite size. However, $\mathbb{Z}$ (the set of all integers) has infinite cardinality, as does $\mathbb{R}$ (the set of all real numbers).

How can we tell if two sets have the same size? The simplest method is to figure out how many elements are in each of the sets. For example, $\{1,2,5\}$ and $\{3,8,11\}$ both have cardinality equal to 3; ergo, these sets have the same size.

However, another way to see if the sets have the same size is to find a one-to-one correspondence between their elements. In other words, instead of counting the elements of the two sets, we show how we can pair each element in the first set with an element in the second. Here is such a one-to-one correspondence for the sets $\{1,2,5\}$ and $\{3,8,11\}$:

> Mathematicians refer to one-to-one correspondences as *bijections*.

$$1 \leftrightarrow 3, \quad 2 \leftrightarrow 8, \quad \text{and} \quad 5 \leftrightarrow 11.$$

This is more trouble than it's worth for small sets that clearly have the same size.

Let's consider a more involved example. Imagine there is a club with seven members (for simplicity, we'll name the members with numbers: $1, 2, 3, \ldots, 7$).

One year, the club has the opportunity to send three of its members to a national conference. There are many different ways to choose the three participants; let A be the set of all possible triples of members:

$$A = \{123, 124, 125, \ldots, 567\}.$$

Each element of A represents a selection of three club members. For example, "237" means members 2, 3, and 7 go to the conference.

The following year, the club's members learn they are permitted to send four members to the national conference. Let B be the set of all quadruples of members:

$$B = \{1234, 1235, 1236, \ldots, 4567\}.$$

Thus A is the set of all triples and B is the set of all foursomes of club members.

Do A and B have the same size?

If you carefully write the two sets out in full, you should find that A and B have the same number of members. Writing down all the possibilities is tedious and prone to errors.

However, in this case it is easier to demonstrate that A and B have the same cardinality by finding a one-to-one correspondence between their elements. Here's the idea:

The club members decide that whoever went to the conference the first year is ineligible to go to the conference in the second; the other four members will go. What we see is that each choice of a threesome for the first year's conference corresponds with the complementary foursome for the next year's meeting. In this way, we can pair every triple with a foursome. For example, if club members 1, 4, and 5 attended the first conference, then 2, 3, 6, and 7 attend the second. We indicate this as $145 \leftrightarrow 2367$.

If we were to write out all the possibilities, the lists would look as shown here in the margin. This one-to-one correspondence between members of A and members of B demonstrates that the two sets have the same size.

> We will see the size of the sets A and B later. The point is, we don't need to know $|A|$ or $|B|$ to determine that $|A| = |B|$.

$123 \leftrightarrow 4567$
$124 \leftrightarrow 3567$
$125 \leftrightarrow 3467$
$\vdots$
$356 \leftrightarrow 1247$
$\vdots$
$567 \leftrightarrow 1234$

You can try writing out the two sets in full and then counting how many elements are in each (but the one-to-one correspondence obviates the need for the tedium). This is done for you on page 85.

In summary, we have two methods to show that a pair of *finite* sets have the same size: we can figure out exactly how many elements are in each or we can find a one-to-one correspondence between their elements. However, if the sets contain infinitely many elements, the first method doesn't apply: there is no number—no integer—that reports, say, the cardinality of $\mathbb{R}$ (the set of real numbers). To show that two infinite sets have the same size, we must produce a one-to-one correspondence between their elements. Here's an example.

Recall that $\mathbb{Z}$ stands for the set of integers:

$$\mathbb{Z} = \{\ldots -3, -2, -1, 0, 1, 2, 3, \ldots\}.$$

Let's use the notation $\mathbb{Z}^+$ to stand for the set of *positive* integers:

$$\mathbb{Z}^+ = \{1, 2, 3, 4, \ldots\}.$$

Do $\mathbb{Z}$ and $\mathbb{Z}^+$ have the same size?

Given the doubt we raised at the beginning of this chapter, you may be tempted to think that $\mathbb{Z}$ has "twice as many" elements as $\mathbb{Z}^+$ and therefore $\mathbb{Z}$ is "twice as infinite." However, these sets have the same size. How do we know? We demonstrate with a one-to-one correspondence.

We make two lists. The first list contains just the positive integers and the second list gives all of the integers (but not in their usual order). By pairing up numbers from the first list with numbers from the second we have the correspondence. This is shown in the box on the next page.

The conclusion is that $\mathbb{Z}^+$ and $\mathbb{Z}$ have the same size. This, perhaps, is not surprising as both are infinite.

$\mathbb{Z}^+$		$\mathbb{Z}$
1	$\leftrightarrow$	0
2	$\leftrightarrow$	1
3	$\leftrightarrow$	-1
4	$\leftrightarrow$	2
5	$\leftrightarrow$	-2
6	$\leftrightarrow$	3
7	$\leftrightarrow$	-3
8	$\leftrightarrow$	4
$\vdots$	$\vdots$	$\vdots$

This chart presents a one-to-one correspondence between $\mathbb{Z}^+$ and $\mathbb{Z}$.

Here are two puzzles for you. The hundredth line of this chart looks like this: 100 $\leftrightarrow$???; what's on the right side? And later in the chart we'll find ??? $\leftrightarrow$ 100 and ??? $\leftrightarrow$ -100; what's on the left side of those lines? Answers appear on page 85.

Infinite sets of unequal size

BY MEANS OF A ONE-TO-ONE CORRESPONDENCE, we showed that $\mathbb{Z}^+$ and $\mathbb{Z}$ have the same size; they are "equally infinite." We are ready for a more interesting question: do $\mathbb{Z}^+$ and $\mathbb{R}$ have the same size? True, both are infinite, but to show they are "equally infinite" we need to find a one-to-one correspondence between their elements. We will show, however, that no such correspondence is possible.

To demonstrate there is a one-to-one correspondence (such as between $\mathbb{Z}^+$ and $\mathbb{Z}$) we show how to pair each element from the first set with an element from the second (and make sure all elements from one set are paired with distinct elements from the other). But how do we show that no such correspondence between $\mathbb{Z}^+$ and $\mathbb{R}$ is possible? We will prove that any attempt to build a chart pairing up elements of $\mathbb{Z}^+$ and $\mathbb{R}$ is doomed to failure because it misses some member of $\mathbb{R}$. Here's how:

Imagine that we've written down a one-to-one correspondence between $\mathbb{Z}^+$ and $\mathbb{R}$. This means we created a chart that looks something like this:

$\mathbb{Z}^+$		$\mathbb{R}$
1	↔	0.349087329190875...
2	↔	3.587908798534216...
3	↔	5.547711170105908...
4	↔	2.224321155332273...
5	↔	9.991260000123015...
⋮	⋮	⋮

Every positive integer appears in the left column and (allegedly) every real number appears in the right. We're now going to prove that no matter how we fill in the right column, there's a real number that we neglected to include.

First, however, we need to pause for a minor technical annoyance. Some real numbers can be written in two different ways using the decimal notation. Consider the number $\frac{1}{4}$. On the one hand, we can write this as 0.25. On the other hand, it is also equal to 0.24999999... where the stream of 9s goes on forever. Both of these are correct decimal representations of $\frac{1}{4}$. The simpler 0.25 can also be written with an infinite stream of 0s as the end. We have

$$\frac{1}{4} = \begin{cases} 0.250000000000\ldots \\ 0.249999999999\ldots \end{cases}$$

It is fine to skip this paragraph on a first reading of this chapter. We're just nailing down a technical detail in order to make the argument complete and unambiguous.

We assume that in the chart, whenever there's a choice of representation, we choose the simpler one that ends in a stream of 0s. The choice doesn't matter for this proof; we just want to be sure we're clear on how we've written the real numbers in the right column.

Now, back to the proof. Imagine we've created a chart that shows an alleged one-to-one correspondence. We're now going to find a real number that's missing from the right column of the chart.

We start by underlining the first digit after the decimal point in the first row, the second digit after

the decimal point in the second row, the third digit in
the third, and so forth. We get something like this:

$\mathbb{Z}^+$		$\mathbb{R}$
1	↔	0.3̲49087329190875...
2	↔	3.5̲87908798534216...
3	↔	5.547̲711170105908...
4	↔	2.2243̲21155332273...
5	↔	9.991260̲000123015...
⋮	⋮	⋮

The sequence of underlined digits is $3, 8, 7, 3, 6, \ldots$.
We use this sequence of digits to build a real number.

We start with a zero and a decimal point: 0.____.
Then we step through the underlined digit sequence
to fill in the places to the right of the decimal point
using these two rules:

(A) If the underlined digit is a 3, append a 7 to the number.

(B) If the underlined digit is not a 3, append a 3 to the number.

Let's work this out for the infinite sequence $3, 8, 7, 3, 6, \ldots$.

We start with 0.____.

The first underlined digit is a 3 and so rule (A)
applies. This tells us to put a 7 after the decimal
point; we now have 0.7____.

The second underlined digit is an 8 and so rule
(B) applies. We tack on a 3 to the end; we now have
0.73____.

Next we have a 7 and rule (B) applies again. We
append another 3: 0.733____.

The fourth underlined digit is another 3 so, by rule
(A), a 7 gets put on the end: 0.7337____.

The fifth digit in the sequence is a 6; by rule (B) we
append a 3 to give 0.73373____.

We continue through the sequence, adjoining
either a 3 or 7 (depending on the current underlined
digit from the sequence) to create a number we name
x. For the sequence in our example, $x = 0.73373\ldots$

78 THE MATHEMATICS LOVER'S COMPANION

and the remaining digits are filled in by applying rules (A) and (B).

Here's a summary of the process:

Underlined digit sequence is:	3	8	7	3	6	...
	↓	↓	↓	↓	↓	
the digits after 0. are:	7	3	3	7	3	...
because of this rule:	A	B	B	A	B	...

The number we produce by this process, x, depends on the chart. With a different chart, we'd build a different x. What we claim is that whatever chart we have, the real number x does *not* appear in the right column and hence the chart is not a valid one-to-one correspondence between $\mathbb{Z}^+$ and $\mathbb{R}$.

Starting at the top, we assert that x is not the real number in the first row of the chart. Here's why: Let's say that first row is $1 \leftrightarrow Y_1$. What is the first digit after the decimal point in Y_1? If it's a 3, then the corresponding digit of x is a 7; but if Y_1's first digit after the point is not a 3, then the corresponding digit in x is a 3. The situation looks like this:

If Y_1 looks like this	Then x looks like this	
0.3yyyyyy...	0.7xxxxxx...	rule (A)
0.?yyyyy...	0.3xxxxxx...	rule (B)

where the ? is any digit other than 3. What we see is that x and Y_1 are different; whatever digit appears after the decimal point in Y_1, the corresponding digit in x is something else. Because x and Y_1 differ in this first decimal place, they are different numbers. So x is not the real number in the first row of the chart.

Does x appear in the second row? Let's call the real number on the right of that row Y_2; we have $2 \leftrightarrow Y_2$. In this case, we inspect the second digit after the decimal point in each of Y_2 and x. If it's a 3 in Y_2, then it's a 7 in x. And if it's not a 3 in Y_2, then it is a 3 in x. Here's the situation:

If Y_2 looks like this	Then x looks like this	
0.y3yyyyy...	0.x7xxxxx...	rule (A)
0.y?yyyyy...	0.x3xxxxx...	rule (B)

where, as before, the ? is any digit other than 3. What we see is that Y_2 and x are different numbers because their second digits after the decimal point are different. Therefore x is not on the second line of the chart.

So far we have shown that x is neither on the first nor the second line of the chart. But if the chart represents a one-to-one correspondence between $\mathbb{Z}^+$ and $\mathbb{R}$, then x has to be *somewhere* in the second column. In other words, x appears on line k of the chart for some positive integer k; that is, we find this line in the chart: $k \leftrightarrow Y_k = x$. But now we run into exactly the same problem as before. What are the kth decimal digits of x and Y_k? If that digit in Y_k is a 3, then the corresponding digit in x is a 7. But if that digit in Y_k is not a 3, then the corresponding digit of x is 3. One last time:

If Y_k looks like this	Then x looks like this	
0.yy...yy3yyyyy...	0.xx...xx7xxxxx...	rule (A)
0.yy...yy?yyyyy...	0.xx...xx3xxxxx...	rule (B)

where the ? is any digit other than a 3. Since the kth digits of x and Y_k disagree, these are different numbers.

What our argument shows is that the real number x does not appear anywhere in the right column of this chart. We can attempt to fix this deficit by making a new chart, perhaps by inserting x at the beginning. However, following the procedure of rules (A) and (B) applied to this new chart, we are guaranteed to find a different number x' that the new chart missed.

The conclusion is that *every* chart is defective! It is impossible to form a one-to-one correspondence between $\mathbb{Z}^+$ and $\mathbb{R}$.

Transfinite numbers

WE HAVE SHOWN THAT $\mathbb{Z}^+$ and $\mathbb{Z}$ have the same size. Not only are both infinite, but we can find a one-to-one correspondence between their elements. The sets $\mathbb{Z}^+$ and $\mathbb{Z}$ have the same (infinite) size.

However, even though $\mathbb{Z}^+$ and $\mathbb{R}$ are both infinite, no such correspondence can be found. Since every positive integer is also a real number, the set $\mathbb{R}$ is demonstrably "bigger" than $\mathbb{Z}^+$. There simply aren't enough positive integers to make a one-to-one pairing with the real numbers.

> The statement that every positive integer is also a real number can also be expressed this way: $\mathbb{Z}^+$ is a *subset* of $\mathbb{R}$.

The size of finite sets can be described with a number. The set $A = \{1, 3, 7, 9\}$ has size four: $|A| = 4$. How do we describe the size of an infinite set? Prior to Cantor's work, the lovely ∞ symbol might have satisfied us. We may be tempted to write $|\mathbb{Z}^+| = \infty$ and $|\mathbb{R}| = \infty$ and then erroneously conclude $|\mathbb{Z}^+| = |\mathbb{R}|$. The ∞ symbol does not provide sufficient nuance to distinguish between the size of $\mathbb{Z}^+$ and the size of $\mathbb{R}$.

To remedy this, Cantor developed an entirely new system of numbers that are beyond the finite. These new numbers are called *transfinite* and they are devised to report the size of infinite sets.

It turns out that $\mathbb{Z}^+$ is a "smallest" infinite set. What does this mean? Let's suppose that X is any infinite set. There might or might not be a one-to-one correspondence between $\mathbb{Z}^+$ and X. But what mathematicians have shown is that there always is a one-to-one correspondence between $\mathbb{Z}^+$ and some *portion* of X. In other words, either $\mathbb{Z}^+$ and X have the same size, or some piece of X (but not its entirety) has the same size as $\mathbb{Z}^+$. Informally, either

$\mathbb{Z}^+$ and X have the same size or else X is larger.

Sets that have the same size as $\mathbb{Z}^+$ are called *countably infinite*. They are the smallest infinite sets. Cantor used the notation $\aleph_0$ to stand for the size of $\mathbb{Z}^+$. In notation, $|\mathbb{Z}^+| = \aleph_0$. Because there is a one-to-one correspondence between $\mathbb{Z}$ and $\mathbb{Z}^+$ we may also write $|\mathbb{Z}| = \aleph_0$. However, since $\mathbb{R}$ is more infinite than $\mathbb{Z}^+$, it is correct to write $|\mathbb{R}| > \aleph_0$. Because $\aleph_0$ represents the size of an infinite set, it is not an integer; rather, it is called a *transfinite* number and $\aleph_0$ is the smallest transfinite number.

The symbol $\aleph$ is the first letter of the Hebrew alphabet: aleph. The notation $\aleph_0$ is typically pronounced "aleph null" or "aleph naught."

To report the size of larger infinite sets, there is an entire universe of transfinite numbers. Sets of size greater than $\aleph_0$ are called *uncountable*, and mathematicians have shown that there is a distinct "level of infinity" one step up from $\aleph_0$. That is, we can show there is a set X with the properties that (a) $|X| > \aleph_0$ but (b) there is no set with size strictly between $|X|$ and $\aleph_0$. Sets of this size are deemed to have cardinality $\aleph_1$. In symbols, $\aleph_0 < \aleph_1$, but there is no transfinite number between $\aleph_0$ and $\aleph_1$.

Indeed, there's an entire staircase of transfinite numbers that begins $\aleph_0 < \aleph_1 < \aleph_2 < \aleph_3 < \cdots$ and so on. This hierarchy extends to transfinite numbers larger than any $\aleph_k$ (where k is an integer). The smallest transfinite number greater than $\aleph_k$ (for all integers k) is called $\aleph_\omega$ and there's infinitely many more beyond that!

Where do the real numbers, $\mathbb{R}$, fit in this scheme? We have established that $|\mathbb{R}| > \aleph_0$. But can we determine the size of $\mathbb{R}$ precisely? How many real numbers are there?

Weirdness in setland

IMAGINE YOU ENTER A BEAUTIFUL BUILDING. A magnificent entryway leads to marble stairs taking you to fantastically appointed rooms. But if you find the door to the basement and descend, the picture quickly changes. There you'll find pipes and wiring, harsh lighting and unfinished floors, and perhaps a few bugs crawling around. The cellar is a scary place, but it's the foundation on which the rest of the structure sits.

This is reasonably apt analogy for the building we call mathematics. As we described at the beginning of this chapter, every concept in mathematics (from integers to circles) is defined in terms of simpler concepts. Necessarily, at some point this process bottoms out and we arrive at the one concept in mathematics from which all others are derived. That concept is the set.

We described sets as *collections of things* but we didn't define *collection* (and it appears to be just another way of saying "set") nor did we worry about what sort of *things* we can collect in sets (since there are no mathematical "things" defined as yet). How do we get out of this pickle?

At first, mathematicians were a bit undisciplined in how they approached this situation. We just assumed that there are things called *sets*, that there is a relation called *is-an-element-of* whose symbol is $\in$, and that we can carry on from there. This got us into a bit of trouble.

> This approach, used by Georg Cantor and others, is known as *naive set theory*.

The "first" set we conceived is the *empty set*. This is a set that has no members and we use the symbol $\emptyset$ to stand for it. The cardinality (size) of the empty set is 0 and the statement "$x \in \emptyset$" is false for any x (since $\emptyset$ has no members).

> Can one set be a member of another? Certainly! For example, the set $\{1,2\}$ is an element of this set:
>
> $$\{0, \{1,2\}, 3, 6, 7\}.$$
>
> This set has five members: the numbers 0, 3, 6, and 7 as well as the set $\{1,2\}$.

Then we thought: We can define sets in terms of the properties of their members. For example, the set of even numbers can be described this way:

$$\{x \mid x \text{ is an integer and } x/2 \text{ is also an integer}\}.$$

In general, the notation $\{x \mid \text{properties of } x\}$ stands for the set of all things whose elements satisfy the listed properties. And this got us into a heap of trouble.

At the start of the twentieth century, the philosopher-mathematician Bertrand Russell thought about this set:

$$A = \{x \mid x \text{ is a set and } x \notin x\}.$$

One of Russell's many accomplishments is a Nobel Prize in Literature, awarded in 1950.

This is the set consisting of those sets that are not members of themselves. For example, the empty set satisfies this property: $\emptyset \notin \emptyset$ because $\emptyset$ doesn't contain any elements. The criteria for a set x to be in A is simple: it must not be a member of itself. Because $\emptyset \notin \emptyset$ we know that $\emptyset \in A$.

And then Russell asked the fatal question: Is $A \in A$?

- If the answer is "yes," then $A \in A$. In order for $A \in A$, it must be the case that A satisfies the defining property: A is not a member of itself. In symbols, $A \notin A$.

- If the answer is "no," then $A \notin A$. But then A satisfies the defining property for A, which means A is a member of A. In symbols, $A \in A$.

If we think $A \in A$, we must conclude $A \notin A$. If we think $A \notin A$, then we must conclude $A \in A$. It cannot be the case that A is both a member and not a member of itself! Something is terribly wrong.

This paradox is called *Russell's antinomy*.

One of the implications of this contradiction is that the set A cannot exist. There is no such set.

In the years following Russell's work, a new approach to set theory was developed. A clear, unambiguous (but rather technical) collection of rules were posited about how sets behave and how they can be formed. The definitions of *set* and $\in$ are implicit in this method. We don't really say what these are; we only describe the properties they have. Then we simply assume there are such things as sets that have the properties delineated by the list of rules and work from there. The new rules prevent Russell's paradox

This new, successful approach to defining sets is known as *axiomatic set theory*. The commonly accepted rules that describe how sets behave and how to form sets are named after the rules' creators, Ernst Zermelo and Abraham Fraenkel: They are known as the ZF Axioms.

from rearing its ugly head, and no further contradictions have arisen.

WE RETURN TO THE QUESTION: HOW MANY REAL NUMBERS ARE THERE? We know that the number of positive integers is $\aleph_0$. And we know that $|\mathbb{R}| > \aleph_0$. But is it the case that $|\mathbb{R}| = \aleph_1$? This would mean that there are no sets that are an intermediate size between $\mathbb{Z}^+$ and $\mathbb{R}$. Cantor believed that $|\mathbb{R}| = \aleph_1$ but could not prove it; he called this conjecture the *continuum hypothesis*. This problem was of tremendous interest. In 1900, the German mathematician David Hilbert developed a list of what he considered to be the twenty-three most important problems for the coming twentieth century. A proof (or disproof) of the continuum hypothesis was first on his list.

Hilbert's first problem has been solved, but not in a way he would have expected. The short, but complete, answer is: "it depends."

Ugh. One of the prized aspects of mathematics is that its problems (usually!) have an unambiguous answer. The "maybe yes/maybe no" solution to the continuum hypothesis seems to belie that definitiveness. How can the answer be so wishy-washy?

The work of Kurt Gödel (in the 1940s) and Paul Cohen (in the 1960s) shows that the standard rules of axiomatic set theory are not rich enough to answer this question. More specifically, they showed that one cannot prove nor disprove that there is a set whose size is strictly between that of $\mathbb{Z}^+$ and $\mathbb{R}$. Here's another way to think about this: One may safely assume that $|\mathbb{R}| = \aleph_1$ or that $|\mathbb{R}| > \aleph_1$. We simply have two different systems of mathematics; both are correct—they're just different.

Here is a complete list of the elements of the sets A and B described on page 73: the triple and quadruples we can make from the numbers 1 through 7. There are 35 elements in each of these sets, and we list them here using the one-to-one correspondence we described earlier:

123 ↔ 4567	124 ↔ 3567	125 ↔ 3467	126 ↔ 3457	127 ↔ 3456
134 ↔ 2567	135 ↔ 2467	136 ↔ 2457	137 ↔ 2456	145 ↔ 2367
146 ↔ 2357	147 ↔ 2356	156 ↔ 2347	157 ↔ 2346	167 ↔ 2345
234 ↔ 1567	235 ↔ 1467	236 ↔ 1457	237 ↔ 1456	245 ↔ 1367
246 ↔ 1357	247 ↔ 1356	256 ↔ 1347	257 ↔ 1346	267 ↔ 1345
345 ↔ 1267	346 ↔ 1257	347 ↔ 1256	356 ↔ 1247	357 ↔ 1246
367 ↔ 1245	456 ↔ 1237	457 ↔ 1236	467 ↔ 1235	567 ↔ 1234

Solution to the puzzle on page 75: The hundredth line reads 100 ↔ 50 and later we find 200 ↔ 100 and 201 ↔ −101.

9
Fibonacci Numbers

Squares and dominoes

WE BEGIN WITH A TILING PROBLEM. Imagine a long, horizontal 1×10 rectangular frame. We want to fill this frame with *squares* of size 1×1 and *dominoes* of size 1×2. Here's a picture of such a tiling:

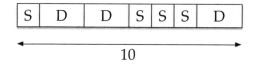

We require the entire 1×10 frame to be completely filled (no gaps) by the S (1×1) and D (1×2) tiles.

The question we ask is: In how many different ways can we tile the 1×10 frame?

It's handy to give a name to the answer to this question: we'll call the answer F_{10} and pretend that "F" stands for "fill."

Drawing all the possible tilings of a 1×10 rectangle and then counting is messy and error prone. A good way to begin is by simplifying the problem.

Instead of trying to find F_{10}, let's begin with F_1. This is almost too easy! We want to count tilings of a 1×1 rectangle by squares (1×1) and dominoes (1×2). Since a domino won't even fit in the space allotted, there's only one solution: pack the frame with a single square. In other words, $F_1 = 1$.

This chapter presents the celebrated Fibonacci numbers 1, 1, 2, 3, 5, 8, 13, 21, and so forth. This sequence is named in honor of Leonardo Bonacci better known as Fibonacci.

Now let's try F_2. The frame is 1×2. We can either fill the frame with two squares or we can fill it with one domino. There are two possible ways to fill a 1×2 frame, and so $F_2 = 2$.

Next: How many ways to tile a 1×3 frame? We can pack three squares (SSS) as one possibility. Or we can put in a single domino (we can't fit two) that is either flush left (DS) or flush right (SD). There are three ways to tile and we have $F_3 = 3$.

One more special case: How many ways can we pack a 1×4 frame with squares and dominoes? Here's a picture containing all the possibilities:

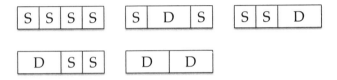

We listed five, but did we find them all? Here's a way to check.

The leftmost tile in the frame may be either a square or a domino. In the figure, the top row shows those tilings that start with an S and the bottom row lists those that start with a D.

When the leftmost tile is an S, we need to fill in the rest of the frame with S's and D's. Well, the rest of the frame has dimensions 1×3 and we already worked out that there are $F_3 = 3$ ways to do that.

Now when the leftmost tile is a D, we again need to fill in the rest of the frame with S's and D's. But in this case, the remaining space to fill has dimensions 1×2. The number of ways to fill a 1×2 frame is $F_2 = 2$.

All told, there are $3 + 2 = 5$ ways to tile the 1×4 frame, and we have confirmed that $F_4 = 5$.

Now it's your turn. Take a few minutes to find all the tilings of a 1×5 frame. There aren't too many and the solution is on page 101. We'll wait!

WELCOME BACK. We hope you found that $F_5 = 8$ because there are 5 tilings that begin with an S and 3 tilings that begin with a D.

Let's summarize our situation. We are using the notation F_n to stand for the number of ways to tile a $1 \times n$ frame with squares and dominoes. The original problem is to work out the value of F_{10} and here are the values we know so far:

F_1	F_2	F_3	F_4	F_5
1	2	3	5	8

Let's keep going. What is F_6? We could take the time to draw all the possibilities, but this is getting tedious. Instead, we break the problem into two pieces. How many tilings of a 1×6 frame are there in which (a) the first tile is an S or (b) the first tile is a D? The great news is that we already know the answers to both of these questions!

In case (a), if the first tile is an S, then the remaining empty space is a 1×5 frame and there are $F_5 = 8$ ways to fill that. In case (b), the remaining empty space is a 1×4 frame and there are $F_4 = 5$ ways to fill that. Therefore $F_6 = F_5 + F_4 = 8 + 5 = 13$.

What's the value of F_7? By the same argument, $F_7 = F_6 + F_5 = 13 + 8 = 21$. What is F_8? It's $F_7 + F_6 = 21 + 13 = 34$. And so on. What we observe is the following relation:

$$F_n = F_{n-1} + F_{n-2}.$$

We're only a few more additions away from obtaining the value of F_{10}. Work this out and check your answer on page 101.

The Fibonacci numbers

THE FIBONACCI NUMBERS are the sequence

$$1, 1, 2, 3, 5, 8, 13, 21, 34, 55, 89, \ldots$$

that is generated by the following rules:

- The first two terms are 1 and 1, and

- each subsequent term is the sum of two that precede it.

We use the notation F_n to stand for the nth Fibonacci number beginning with $n = 0$:

$$F_0 = 1, \quad F_1 = 1, \quad F_2 = 2, \quad F_3 = 3, \quad F_4 = 5,$$

and so on. We generate the rest of the sequence by repeated application of the rule

$$F_n = F_{n-1} + F_{n-2}.$$

In other words, the Fibonacci numbers are precisely the solution to the tiling problem we described at the beginning of this chapter.

THE FIBONACCI NUMBERS enjoy a myriad of interesting properties that have fascinated mathematicians for ages. Many authors have written on this subject; our modest goal in this chapter is to illustrate some interesting methods of mathematical proof and then to close with a formula for the nth Fibonacci number.

In the tiling problem, $F_1 = 1$ and $F_2 = 2$. But here we begin the sequence with two 1s. To reconcile this with the previous section we let $F_0 = 1$. Does this make sense with the tiling problem? The question when $n = 0$ is this: In how many ways can we tile a 0×1 frame with squares and dominoes? Of course, we can't fit *any* tiles into this frame, so one might be tempted to say the answer is 0, but that's not right. There is a way to completely "fill" a 0×1 rectangle; it already has no empty space! So we don't need to put any tiles into the frame to fill it. Therefore there is exactly one way to fill the frame, and that one way is to not put any tiles into it.

Does this make sense to you? If so, congratulations. You have the soul of a mathematician!

Sums of Fibonacci numbers

WHAT HAPPENS WHEN WE ADD the first several Fibonacci numbers? That is, what can we say about the sum $F_0 + F_1 + \cdots + F_n$ for various values of n. Let's do some calculations and see what's going on.

Look closely at the results (right-hand sides) of the calculations in the box on the next page. Do you see a pattern? Take a few minutes before reading on as it's better to discover this result yourself than to simply read the answer.

DID YOU FIND THE PATTERN? If not, go back and keep looking!

$$1 = 1$$
$$1 + 1 = 2$$
$$1 + 1 + 2 = 4$$
$$1 + 1 + 2 + 3 = 7$$
$$1 + 1 + 2 + 3 + 5 = 12$$
$$1 + 1 + 2 + 3 + 5 + 8 = 20$$
$$1 + 1 + 2 + 3 + 5 + 8 + 13 = 33$$
$$1 + 1 + 2 + 3 + 5 + 8 + 13 + 21 = 54$$
$$1 + 1 + 2 + 3 + 5 + 8 + 13 + 21 + 34 = 88$$
$$1 + 1 + 2 + 3 + 5 + 8 + 13 + 21 + 34 + 55 = 143$$
$$1 + 1 + 2 + 3 + 5 + 8 + 13 + 21 + 34 + 55 + 89 = 232$$

Sums of Fibonacci numbers.

What we hope you found is that the numbers on the right are all one less than a Fibonacci number. For example, adding F_0 through F_5 gives

$$F_0 + F_1 + F_2 + F_3 + F_4 + F_5 = 1 + 1 + 2 + 3 + 5 + 8 = 20 = F_7 - 1.$$

Adding F_0 through F_6 gives 33, which is one less than $F_8 = 34$. This pattern holds for all of the examples we presented and so it is reasonable to conjecture that the following formula works for all nonnegative integers n:

$$F_0 + F_1 + F_2 + \cdots + F_n = F_{n+2} - 1. \qquad (*)$$

Perhaps seeing the equation $(*)$ work nearly a dozen times is enough to convince you that the pattern holds for all n, but mathematicians want proof. We are pleased to present to you two rather different arguments that $(*)$ holds for all integers n. The techniques are called *proof by induction* and *combinatorial proof*.

Proof by induction

EQUATION (∗) IS ACTUALLY INFINITELY MANY EQUATIONS. To show that (∗) holds for a specific value of n, say $n = 6$, is just a matter of arithmetic. We look up the values of F_0 through F_6 and add:

$$F_0 + F_1 + \cdots + F_6 = 1 + 1 + 2 + 3 + 5 + 8 + 13 = 33$$

and then we look up F_8 (it's 34) and check that we got the expected answer (we did).

Next we check $n = 7$. We could start over by adding F_0 through F_7, but that seems like a waste of time. We already added through F_6, so let's take advantage of that. Since $F_7 = 21$, we just do this:

$$(F_0 + F_1 + \cdots + F_6) + F_7 = 33 + 21 = 54.$$

As before, since $F_9 = 55$ we see that the result is $F_9 - 1$ as we expect.

Let's show that (∗) works for $n = 8$ and we'll be even lazier. Here's what we already know and what we want to calculate:

> We know: $F_0 + F_1 + \cdots + F_7 = F_9 - 1$
> We want: $F_0 + F_1 + \cdots + F_7 + F_8 = ?$

Let's use what we know to help us with what we want, like this:

$$(F_0 + F_1 + \cdots + F_7) + F_8.$$

We know the value of the sum inside the parentheses: it's $F_9 - 1$ so let's drop that in:

$$(F_0 + F_1 + \cdots + F_7) + F_8 = (F_9 - 1) + F_8$$

So now we just have to evaluate $F_9 - 1 + F_8$. We could do that by looking up the values of F_8 and

F_9 and then doing some arithmetic. But we can be even lazier! We know that $F_8 + F_9 = F_{10}$, and so we complete our calculation without ever having to actually calculate (or even looking up any Fibonacci numbers):

$$(F_0 + F_1 + \cdots + F_7) + F_8 = (F_9 - 1) + F_8$$
$$= (F_8 + F_9) - 1 = F_{10} - 1.$$

We have successfully verified the case $n = 8$ by resting on the work we already did in the case $n = 7$.

The $n = 9$ case can be derived from the $n = 8$ case by the same trick. (Try it yourself!) Indeed, given that we've verified $(*)$ for one value of n we can use that work to quickly verify the next case for $(*)$.

WE ARE NOW READY TO GIVE A COMPLETE PROOF of equation $(*)$. As we mentioned, $(*)$ is not a single equation but an infinite list of equations: one for each value of n from 0 on up. Here's how the proof works.

We first prove $(*)$ for the easiest case, $n = 0$. That is, we just have to show that F_0 is equal to $F_{0+2} - 1$. Since $F_0 = 1$ and $F_2 = 2$ we clearly have $F_0 = F_2 - 1$. (That's just a fancy way of writing $1 = 2 - 1$.)

Next we show how if we have already proven $(*)$ for one value of n (say $n = k$) then we can by "autopilot" prove $(*)$ for the next value of n (say $n = k + 1$). All we need to do is to show how that "autopilot" works. This is what we need to do:

Let k be some specific number.

| Suppose that we know this: | $F_0 + F_1 + \cdots + F_k = F_{k+2} - 1$ |
| We seek the value of this: | $F_0 + F_1 + \cdots + F_k + F_{k+1} = ?$ |

We want to evaluate $F_0 + F_1 + \cdots + F_k + F_{k+1}$ and we already know the sum of the terms up to F_k. Let's put in what we know:

$$(F_0 + F_1 + \cdots + F_k) + F_{k+1} = (F_{k+2} - 1) + F_{k+1}.$$

The right-hand side is $F_{k+2} - 1 + F_{k+1}$ and we know how to add consecutive Fibonacci numbers. We have

$$F_{k+2} - 1 + F_{k+1} = (F_{k+1} + F_{k+2}) - 1 = F_{k+3} - 1.$$

Dropping that in we conclude

$$(F_0 + F_1 + \cdots + F_k) + F_{k+1} = F_{k+3} - 1.$$

Let's be clear on what we just did. If we already know that $(*)$ is correct when we sum up to F_k, then $(*)$ must also be correct when we add up to F_{k+1}.

Summarizing:

- Equation $(*)$ is true for $n = 0$.

- If one version of $(*)$ is known to be true, the next one in line must also be true.

We can now say with certainty that $(*)$ is true for all values of n. Is $(*)$ true when $n = 4987$? We can be sure if it's true for $n = 4986$, which in turn rests on the case $n = 4985$, which rests on the case $n = 4984$, and so forth, all the way back to the case $n = 0$. In this way, equation $(*)$ is true for all possible values of n.

This method of proving $(*)$ is known as *mathematical induction* (or *proof by induction*). It is a technique in which a base case is checked and then a template proof is given that shows how each subsequent case follows from an already proven case.

Combinatorial proof

WE NOW PRESENT A TOTALLY DIFFERENT PROOF OF $(*)$ ALTOGETHER. The central idea in this second proof is to use the fact that F_n is the answer to a counting problem: the number of tilings of a $1 \times n$ frame with squares and dominoes.

For convenience, here is the equation we want to prove:

$$F_0 + F_1 + F_2 + \cdots + F_n = F_{n+2} - 1. \qquad (*)$$

The idea is to understand both sides of this equation as the solution to a counting problem. If we can demonstrate that the left and right sides of $(*)$ are correct answers to the same counting question, then they must be the same. This technique is known as *combinatorial proof*.

The problem is: What is the counting question to which $(*)$ gives two correct answers? This conundrum is like the television show *Jeopardy!* in which contestants must formulate questions when challenged with their answers.

The right-hand side is easier to think about, so let's begin there. The answer is: $F_{n+2} - 1$. What's the question? If the answer were simply F_{n+2} then we have a question ready to use: How many tilings are there of an $(n+2)$-long frame using squares and dominoes?

That question is nearly what we want, but it's off by one. By gently modifying this question we can adjust the answer by 1. That is, let's discard one of the tilings and count the rest. The tricky part is to think about one of the tilings as being "different" in some way than all the others. Is one of these tilings unlike any of the others?

Every tiling is made up of squares and dominoes. There is only one tiling that is entirely squares; all the rest have at least one domino. Let's use that as the basis of our question:

> **Question**: How many tilings are there of an $(n+2)$-by-1 frame using squares and dominoes that have at least one domino?

We'll now present two correct answers to the Question. Since they are both correct answers to the same question, they must be equal.

THE FIRST ANSWER TO THE QUESTION is one we

The word combinatorial *is the adjective form of the noun* combinatorics, *which is the branch of mathematics rooted in counting problems such as the tiling problem from the beginning of this chapter. The word* combinatorics *derives from the word* combinations.

already discussed. There are F_{n+2} tilings of an $(n+2)$-frame. Among these, there is only one that doesn't have any dominoes: it's the tiling entirely composed of squares.

Therefore, Answer #1 to the Question is $F_{n+2} - 1$.

THE SECOND ANSWER TO THE QUESTION gives—we hope—the left-hand side of $(*)$. Let's see how this works.

We need to count tilings that include at least one domino. So let's think about where in the tiling the first (leftmost) domino occurs. There are $n + 2$ positions in the frame and the leftmost domino can be in position 1, 2, 3, or any position up to $n + 1$, but it can't start in position $n + 2$.

Let's look at the case $n = 4$. We seek tilings of a 1×6 frame with at least one domino. We know the answer is $F_6 - 1 = 13 - 1 = 12$, but we'll work out this answer another way.

The leftmost domino may be in any one of positions 1 through 5 as shown in this picture:

| D | S | S | S | S | | S | D | S | S | S | | S | S | D | S | S | | S | S | S | D | S | | S | S | S | S | D |

| D | D | S | S | | S | D | D | S | | S | S | D | D |

| D | D | D | | S | D | S | D |

| D | S | D | S |

| D | S | S | D |

The left column in the figure shows those tilings of a 1×6 frame in which the leftmost domino is in the first position, the second column shows those in which the leftmost domino is in the second position, and so on.

How many tilings are in each column?

There are five in the first column and, when we ignore the leftmost domino, we see exactly the $F_4 = 5$ tilings of a 1×4 frame.

In the second column we find three tilings. Ignore

the initial square and leftmost domino; what remains are the $F_3 = 3$ tilings of a 1×3 frame.

Likewise for the remaining columns. When we count how many tilings have their leftmost domino in a given position, all the variability happens to the right of that first domino. Here's what we find:

Location of first domino	Number of tilings
1	$5 = F_4$
2	$3 = F_3$
3	$2 = F_2$
4	$1 = F_1$
5	$1 = F_0$

Therefore the number of tilings of a 6-long frame with at least one domino is $F_4 + F_3 + F_2 + F_1 + F_0 = 12$. Conclusion:

$$F_0 + F_1 + F_2 + F_3 + F_4 = 12 = F_6 - 1.$$

LET'S DIVE INTO THE GENERAL CASE. We're given that the frame is $n+2$-long. How many tilings have their leftmost domino in position k? Since the leftmost domino is in position k the first $k-1$ slots are filled with squares and the next two slots are covered by the domino. These $(k-1) + 2 = k+1$ positions allow no flexibility. However, the remaining $(n+2) - (k+1)$ slots can be tiled in any way we choose. Since $(n+2) - (k+1) = n-k+1$, the number of ways to tile this portion of the frame is F_{n-k+1}. This is illustrated in the following diagram:

Here, k can be any value from 1 to $n+1$; we can't have $k = n+2$ because then the domino would extend beyond the end of the frame.

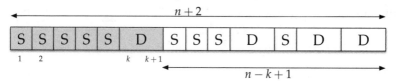

Notice that the first domino in the $n+2$-frame is at position k. That means that positions 1 through $k+1$ are set. The remaining $n-k+1$ positions can be tiled in F_{n-k+1} ways where k ranges from 1 to $n+1$.

Notice that as k varies from 1 up to $n+1$, the value of $n-k+1$ goes from n down to 0.

Therefore, the number of ways to tile an $n+2$-long frame in which there must be at least one domino is

$$F_n + F_{n-1} + F_{n-2} + \cdots + F_1 + F_0,$$

which is the left-hand side of ($*$) (just written in opposite order).

Therefore, $F_0 + F_1 + \cdots + F_n$ is Answer #2 to the Question.

WE HAVE ASKED THE QUESTION and found two correct answers. Answer #1 is $F_{n+2} - 1$ and Answer #2 is $F_0 + F_1 + \cdots + F_n$. Since these two expressions answer the same counting question, they are equal and ($*$) is proved.

Ratios of Fibonacci numbers and the golden mean

ADDING TWO CONSECUTIVE FIBONACCI NUMBERS yields the following Fibonacci number. In this section we ponder the question: What happens when we divide consecutive Fibonacci numbers. Specifically, we consider the fractions F_{k+1}/F_k for larger and larger values of k. The chart on the following page shows these results from F_1/F_0 through to F_{20}/F_{19}.

As the Fibonacci numbers get larger and larger, the quotients we calculate rapidly approach a constant value that is approximately 1.61803. What is this number?

This number, you may be surprised to learn, is sufficiently famous that if you type it into Google, you are led to a plethora of pages about the *golden mean*. What is this number?

Some call this value the golden ratio.

The ratios of consecutive Fibonacci numbers are not all the same. They are nearly the same, however, for large Fibonacci numbers. So, for the sake of determining what this number 1.61803 is, let's pretend for

Ratio	Fraction	Decimal
F_1/F_0	1/1	1.0
F_2/F_1	2/1	2.0
F_3/F_2	3/2	1.5
F_4/F_3	5/3	1.6666666666666667
F_5/F_4	8/5	1.6
F_6/F_5	13/8	1.625
F_7/F_6	21/13	1.6153846153846154
F_8/F_7	34/21	1.619047619047619
F_9/F_8	55/34	1.6176470588235294
F_{10}/F_9	89/55	1.6181818181818182
F_{11}/F_{10}	144/89	1.6179775280898876
F_{12}/F_{11}	233/144	1.6180555555555556
F_{13}/F_{12}	377/233	1.6180257510729614
F_{14}/F_{13}	610/377	1.6180371352785146
F_{15}/F_{14}	987/610	1.618032786885246
F_{16}/F_{15}	1597/987	1.618034447821682
F_{17}/F_{16}	2584/1597	1.6180338134001253
F_{18}/F_{17}	4181/2584	1.618034055727554
F_{19}/F_{18}	6765/4181	1.6180339631667064
F_{20}/F_{19}	10946/6765	1.6180339985218033

Ratios of consecutive Fibonacci numbers.

a moment that the ratios are all the same. Let's call that ratio x and so we have

$$x = \frac{F_{k+1}}{F_k} = \frac{F_{k+2}}{F_{k+1}} = \frac{F_{k+3}}{F_{k+2}} = \cdots.$$

This means that $F_{k+1} = xF_k$, $F_{k+2} = xF_{k+1}$, and so on. This can be combined like this:

$$F_{k+2} = xF_{k+1} = x^2 F_k.$$

We also know that $F_{k+2} = F_{k+1} + F_k$. Therefore

$$x^2 F_k = xF_k + F_k,$$

and if we divide by F_k and rearrange we have this quadratic equation:

$$x^2 - x - 1 = 0.$$

Using the quadratic formula, we find this equation has two solutions:

$$x = \frac{1+\sqrt{5}}{2} \approx 1.61803 \quad \text{and} \quad x = \frac{1-\sqrt{5}}{2} \approx -0.61803.$$

The first of these is the number we see. The Greek letter φ (phi) is often used to represent the golden mean:

$$\varphi = \frac{1+\sqrt{5}}{2} = 1.618033988749895\ldots$$

What we have observed is that the ratios of consecutive Fibonacci numbers get closer and closer (converge) to φ. This is nice and it also gives us an interesting way to approximate the Fibonacci numbers.

The real sequence of Fibonacci numbers is $F_0, F_1, F_2, F_3, \ldots$. If the ratios F_{k+1}/F_k were all the same, we would have a formula of the form

$$F_n = c\varphi^n$$

where c is some number. To see how well this might work, let's compare F_n and φ^n for several values of n:

F_n	φ^n	F_n/φ^n
1	$\varphi^1 = 1.618034$	0.6180339887498948
2	$\varphi^2 = 2.618034$	0.7639320225002103
3	$\varphi^3 = 4.236068$	0.7082039324993691
5	$\varphi^4 = 6.854102$	0.7294901687515772
8	$\varphi^5 = 11.09017$	0.7213595499957939
13	$\varphi^6 = 17.944272$	0.7244651700109358
21	$\varphi^7 = 29.034442$	0.7232789287212935
34	$\varphi^8 = 46.978714$	0.7237320325750782
55	$\varphi^9 = 76.013156$	0.7235589623033660
89	$\varphi^{10} = 122.991869$	0.7236250692647179
144	$\varphi^{11} = 199.005025$	0.7235998186523744
233	$\varphi^{12} = 321.996894$	0.7236094635280528
377	$\varphi^{13} = 521.001919$	0.7236057795133609
610	$\varphi^{14} = 842.998814$	0.7236071866817582

For larger values of n we find that $F_n/\varphi^n \approx 0.723607$. With a good deal more work, we can demonstrate that the value 0.723607... is given exactly by $\varphi/\sqrt{5}$. In other words, we have

$$F_n \approx \frac{\varphi^{n+1}}{\sqrt{5}}.$$

How good is this approximation? Time for another chart!

F_n	$\varphi^{n+1}/\sqrt{n}$	Rounded
1	0.723607	1.0
1	1.17082	1.0
2	1.894427	2.0
3	3.065248	3.0
5	4.959675	5.0
8	8.024922	8.0
13	12.984597	13.0
21	21.009519	21.0
34	33.994117	34.0
55	55.003636	55.0
89	88.997753	89.0
144	144.001389	144.0
233	232.999142	233.0
377	377.000531	377.0
610	609.999672	610.0
987	987.000203	987.0

Notice that when we round $\varphi^{n+1}/\sqrt{5}$ to the nearest integer, the result is exactly F_n.

If you are bothered by the round-to-the-nearest-integer aspect of this formula, the following formula attributed to Jacques Binet gives an exact result:

$$F_n = \frac{\varphi^{n+1} - (-1/\varphi)^{n+1}}{\sqrt{5}}.$$

Tilings of a 1 × 5 frame
Here are all the tilings of a 1 × 5 frame:

S S S S S	D S S S
S S D S	D S D
S S S D	D D S
S D S S	
S D D	

Notice there are $F_4 = 5$ tilings that begin with an S and $F_3 = 3$ tilings that begin with a D. Therefore there are $5 + 3$ tilings of a 1 × 5 frame, so $F_5 = 8$.

The value of F_{10} (and the answer to the original tiling problem) is 89.

10
Factorial!

Books on the shelf

IN HOW MANY WAYS CAN YOU ARRANGE your books on a shelf? This depends, of course, on how many books you own. Let's begin with a modest example. Suppose your collection contains just three books with the simple titles A, B, and C.

We first choose which book to put at the far left; let's say its A. With that decision in place, there are two possible ways to complete the shelf: either ABC or ACB. So there are two arrangements in which A is the leftmost book.

Or we might decide to put B at the far left. In this case the possible arrangements are BAC or BCA. Or we might put C leftmost, yielding two more arrangements: CAB or CBA.

In total, there are six possible ways to arrange the books:

    ABC   ACB   BAC   BCA   CAB   CBA

Now imagine we acquire another book: D. How many ways can we arrange these four books on our shelf? The approach we used for three books can be used again. We begin by considering which book will be to the far left; let's say it's A. The remaining three books (B, C, and D) fill in the rest of the shelf in 6 possible ways—we know it's 6 because we already figured out that there are 6 ways to arrange three books.

Likewise if B is leftmost: there are 6 ways to arrange the remaining books (A, C, and D). Six again if C is leftmost and 6 again for D. In total, there are $6 \times 4 = 24$ arrangements. Here they are:

ABCD	ABDC	ACBD	ACDB	ADBC	ADCB
BACD	BADC	BCAD	BCDA	BDAC	BDCA
CABD	CADB	CBAD	CBDA	CDAB	CDBA
DABC	DACB	DBAC	DBCA	DCAB	DCBA

Before we take the leap to an arbitrary number of books, let's consider the case of five books: A, B, C, D, and E. As in the previous analyses, we think about which book is leftmost. If it is A, then the remaining four books (B through D) are arranged to A's right. In how many ways? That's exactly the problem we considered before: 24. Similarly if B is leftmost, the other books can be arranged in 24 ways. And likewise when C, D, and E are at the far left. This means, in total, there are $24 + 24 + 24 + 24 + 24 = 5 \times 24 = 120$ ways to arrange the five books.

Here is another way to think about the five-book problem. There are five separate cases to consider based on which book is at the far left. Once that book has been shelved, there are four remaining books to be shelved. So the answer to the five-book problem is 5 times the answer to the four-book problem. A bit of notation will make this easier to write.

Let A_5 denote the answer to the five-book problem. That is, A_5 is the number of ways to arrange five books on a shelf. By considering which book is at the far left, we have the equation

$$A_5 = 5 \times A_4$$

where A_4 stands for the answer to the four-book problem.

Now A_4 can be solved in a similar way. There are four books that might be leftmost; for each choice of

the left book, we have to solve a three-book problem. Therefore
$$A_4 = 4 \times A_3.$$
Our analysis of A_3 reveals that $A_3 = 3 \times A_2$. Even the (very easy) two-book problem succumbs to this analysis: $A_2 = 2 \times A_1$ where, of course, $A_1 = 1$.

Putting this all together we find this:
$$\begin{aligned} A_5 &= 5 \times A_4 \\ &= 5 \times (4 \times A_3) \\ &= 5 \times 4 \times (3 \times A_2) \\ &= 5 \times 4 \times 3 \times (2 \times A_1) \\ &= 5 \times 4 \times 3 \times 2 \times 1 = 120. \end{aligned}$$

The general case is now clear. The number of ways to arrange N books on a shelf is
$$N \times (N-1) \times (N-2) \times \cdots \times 3 \times 2 \times 1. \quad (A)$$

The expression in (A) is known as N *factorial*. Factorial is denoted with an exclamation point, like this: $N!$. For example, $6! = 6 \times 5 \times 4 \times 3 \times 2 \times 1 = 720$.

Is there a formula?

WHAT IS THE VALUE OF $10!$? That's simply a matter of multiplying together the numbers 1 through 10 to find
$$10! = 10 \times 9 \times 8 \times \cdots \times 3 \times 2 \times 1 = 3{,}628{,}800.$$

To calculate $20!$ we need to multiply twenty numbers. Calculating $100!$ is a lot of work. Is there a quick way to the result?

Here's an aesthetically pleasing but computationally worthless idea. To calculate $10!$ "all" we need to do is calculate $9!$ and then multiply that result by 10. That's because
$$10! = 10 \times [9 \times 8 \times \cdots \times 3 \times 2 \times 1] = 10 \times 9!.$$

For readers who have studied calculus, here is another aesthetically pleasing formula:
$$N! = \int_0^\infty x^N e^{-x}\, dx.$$
This formula is not useful for calculation, but it allows us to do the devilish maneuver of substituting $N = \frac{1}{2}$ leading to $\frac{1}{2}! = \sqrt{\pi}/2$.

For an arbitrary value N we have this generalization:

$$N! = N \times [(N-1) \times (N-2) \times \cdots \times 3 \times 2 \times 1]$$

yielding this formula:

$$N! = N \times (N-1)! \qquad \text{(B)}$$

Formula (B) is lovely, but it's not of much help for calculating 20!. It tells us to first calculate 19! and multiply that result by 20. And, indeed, it tells us how we should calculate 19!: first calculate 18! and then multiply by 19. In the end, it's a clever way to disguise the fact that we are going to end up multiplying together the numbers 1 through 20.

WE WERE HOPING FOR A SHORT CUT. Is there any reason to think we can speed things up a bit? We can draw some inspiration by considering *triangular numbers*: numbers of the form

$$1 + 2 + 3 + \cdots + N.$$

For example, the fifth triangular number is $1 + 2 + 3 + 4 + 5 = 15$. Let's write T_N to stand for the Nth triangular number defined to be

$$T_N = N + (N-1) + (N-2) + \cdots + 3 + 2 + 1.$$

For example,

$$T_{10} = 10 + 9 + 8 + 7 + 6 + 5 + 4 + 3 + 2 + 1 = 55.$$

This looks just like factorial, but with addition in place of multiplication. Is there a way to calculate T_{10} without having to add ten numbers?

The delightful answer is yes and the reasoning is simple and elegant. We can write the sum for T_{10} with the terms in ascending order and in descending order like this:

```
 1  +  2  +  3  +  4  +  5  +  6  +  7  +  8  +  9  + 10
10  +  9  +  8  +  7  +  6  +  5  +  4  +  3  +  2  +  1
```

These numbers are called *triangular* because they count the number of circles in a diagram like this:

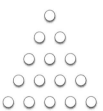

If we were to add up all these 20 numbers, the result would be twice the value of T_{10}. But rather than add these numbers horizontally, let's first add them *vertically*:

$$
\begin{array}{ccccccccccccccccccccc}
1 & + & 2 & + & 3 & + & 4 & + & 5 & + & 6 & + & 7 & + & 8 & + & 9 & + & 10 \\
10 & + & 9 & + & 8 & + & 7 & + & 6 & + & 5 & + & 4 & + & 3 & + & 2 & + & 1 \\
\hline
11 & + & 11 & + & 11 & + & 11 & + & 11 & + & 11 & + & 11 & + & 11 & + & 11 & + & 11
\end{array}
$$

The bottom row has ten terms all equal to 11, so that sum is simply $10 \times 11 = 110$. But since this is double the value of T_{10}, we have that $T_{10} = 110 \div 2 = 55$.

You can check this with a calculator in less than a minute! Add all the numbers from 1 to 10 and confirm that the sum is 55.

In general, we can calculate like this. To find T_N we write the numbers from 1 to N ascending and descending, and then we add vertically:

$$
\begin{array}{ccccccccccccc}
1 & + & 2 & + & 3 & + & \cdots & + & (N-2) & + & (N-1) & + & N \\
N & + & (N-1) & + & (N-2) & + & \cdots & + & 3 & + & 2 & + & 1 \\
\hline
(N+1) & + & (N+1) & + & (N+1) & + & \cdots & + & (N+1) & + & (N+1) & + & (N+1)
\end{array}
$$

The bottom row has N terms, each equal to $N+1$; thus the sum of these numbers is $N \times (N+1)$. Since this is a "double portion" of T_N, we conclude that

$$T_N = \frac{N \times (N+1)}{2}.$$

To calculate T_{100} we do not have to add one hundred numbers. We just have to calculate

$$100 \times 101 \div 2 = 5050$$

and we have our answer.

IS THERE A SIMILAR, ELEGANT FORMULA to calculate factorials? Sadly, there is not. We are, however, fortunate to have a good approximate formula due to James Stirling. Stirling's approximation is

James Stirling was a Scottish mathematician who lived in the 1700s.

$$N! \approx \sqrt{2\pi N} \left(\frac{N}{e}\right)^N. \qquad (C)$$

This formula involves two remarkable numbers discussed in other chapters: $\pi \approx 3.14159$ is the ratio of a circle's circumference to its radius (Chapter 6) and $e \approx 2.71828$ is Euler's number (Chapter 7).

Stirling's formula has better accuracy with larger values of N. For example, for $N = 10$ we have $10! = 3{,}628{,}800$ and formula (C) gives $3{,}598{,}695.6187$, which is only off by about 0.8%.

For $N = 20$, we have these results:

$$20! = 2432902008176640000$$
$$\text{formula (C)} = 2422786846761133393.6839075390,$$

which is off by about 0.4%. If we jump to $N = 1000$, the relative error between 1000! and formula (C) is less than 0.01%.

A puzzle

THE NUMBER 145 IS CALLED a *factorion* because it has the following amusing property. If we sum the factorials of its digits, we get the original number:

$$1! + 4! + 5! = 1 + 24 + 120 = 145.$$

The numbers 1 and 2 are also factorions (but zero is not—see below). There is only one other factorion. See if you can find it.

This is tough to do without writing a computer program. The answer is on the next page.

What is 0!?

MANY PEOPLE HAVE AN OVERWHELMING URGE to answer: 0! is zero! (The second exclamation point is for emphasis.) This appears to make sense since the first factor in $N!$ is N, and anything multiplied by zero must be zero. However, mathematicians define 0! to be equal to 1, and we close this chapter with a few words of explanation.

In Chapter 1 we encountered the concept of an *empty product*—a multiplication problem with no terms (see page 10). Zero factorial is an instance of an empty product. For any N, we note that $N!$ has exactly N terms. This is clear for positive N, but it is also true when $N = 0$. In a sense, the definition of $N!$ is that we multiply all integers from 1 to N together. In case $N = 0$ there are no such numbers, and so the product is empty. By convention, empty products evaluate to 1.

Here's another rationale for defining 0! equal to 1. If we substitute $N = 1$ into formula (B) we have

$$N! = N \times (N-1)! \quad \Longrightarrow \quad 1! = 1 \times 0!$$

and since $1! = 1$, we must have 0! equal to 1.

Finally, let's return to our bookshelf. In how many ways can we arrange N books on the shelf if we have no books? There's one and only one arrangement available to us: leave the shelf empty.

Just like the number 145, the number 40,585 is a factorion because it equals the sum of the factorials of its digits:

$$4! + 0! + 5! + 8! + 5! = 24 + 1 + 120 + 40320 + 120 = 40585.$$

11
Benford's Law

WE HOLD THIS TRUTH TO BE SELF-EVIDENT that all digits are created equal. No, we don't mean "equal to each other"—clearly not! What we mean is that when we consider the ten decimal digits, 0 through 9, we expect all of them to get equal play in the world of numbers.

The sad truth is that digits can be as vain as humans; they all want to be first in line. Indeed, we call the leftmost digit in a number the *most significant digit*—here's why: Imagine you are about to purchase an item the costs $43.52. Which of these digits is most important to you? The 4 carries the most importance while the 2 at the end hardly matters. You would care quite a bit if the 4 were suddenly changed to a 9, but you don't have the same feeling about changing the 2 at the end.

Those who expect the universe to be fair might expect that the digits each get equal shares playing the role of most significant, but alas poor zero! It doesn't get to be the most significant digit; that honor is only held by the other nine digits. They all want to be the most significant as often as possible.

We would expect the digits 1 though 9 to share nicely and each ought to begin one-ninth (about 11%) of the numbers. Surely there can't be "more" numbers that start with a 2 than that start with 5.

> For a number such as 0.053, the 5 is considered to be the most significant digit. When we speak of the most significant digit, it is always the first nonzero digit.

Right?

Wild measurements

THE ASSERTION THAT THE NUMERALS 1 through 9 are equally represented as the most significant digit of a number makes sense when we think about a range of numbers, say from 1 to 999,999. In this case, each of 1 through 9 appears equally often as the leading digit.

However, this outcome is biased by the range from which we chose the numbers. If instead we looked at the numbers, say, from 1 to 19, then each of 2 through 9 appears just once in the leftmost position while 1 is the most significant digit for the other eleven values.

To be fair let's collect numbers from the real world. We need to be a bit careful not to pick measurements that are likely to be concentrated in a narrow range. So let's not collect "heights of adults" because when expressed in centimeters, nearly all of the measurements begin with a 1 or a 2 (because there are few adults whose height is greater than 299 cm or less than 100 cm).

If the heights-of-adults measurements are expressed in feet then nearly all of the values will begin with one of 4, 5, or 6.

To make sure that all digits have a "fair" chance to land in the leftmost position, let's collect measurements over a wide range of values. For example, consider the populations of nations. These values range from over a billion (for China and India) to fewer than ten thousand (for the island nation of Nauru). In addition to populations we collected the following measurements for hundreds of nations:

Our source for these populations and the other measurements is the CIA World Fact Book, available on the web.

- the gross domestic product (in U.S. dollars),
- the number of airports,
- the total area (in square kilometers),
- the annual electricity production (in kilowatt-hours),

- the annual consumption of refined petroleum products (in barrels),
- the total length of all roads (in kilometers),
- the combined length of all railroads (in kilometers), and
- the number of telephones.

In this way, we collected nearly two thousand measurements and then tallied how many begin with a 1, how many begin with a 2, and so forth. Here's the result:

First digit	Count	Percentage
1	582	29.7%
2	338	17.2%
3	236	12.0%
4	206	10.5%
5	173	8.8%
6	123	6.3%
7	112	5.7%
8	101	5.2%
9	90	4.6%

Amazingly, the most frequent first digit is 1 (about 30%) and the least common is 9 (less than 5%)!

We encourage our readers to repeat this experiment on their own. Using an almanac or other reference materials, collect first digits from lengths of rivers, heights of mountains, stock prices, average weights of different species of animals, number of words in novels, rice production by country, and so forth.

Collect enough measurements that span a wide range of values, and you should see the same pattern. The leftmost digit of these numbers is most often a 1, and the frequencies trail off with 9 being the least frequent.

THE INEQUITABLE DISTRIBUTION OF INITIAL DIGITS is known as *Benford's Law*. This name honors

Frank Benford, who published an article about this phenomenon in 1938, although it was reported earlier by Simon Newcomb in 1881.

Benford's law is more specific than "1 appears most and 9 least" as the most significant digit. Benford's law asserts that—given enough data—the frequencies are as follows:

Digit	Frequency
1	30.10%
2	17.61%
3	12.49%
4	9.69%
5	7.92%
6	6.69%
7	5.80%
8	5.12%
9	4.58%

These numbers are approximations. The predicted frequency for 1 in the leftmost position is 30.102999566398114...%. We'll explain later from whence these values arise.

Multiplication tables

THERE'S ANOTHER PLACE WHERE WE CAN FIND an inequitable distribution of initial digits, and that's in a standard multiplication table:

×	1	2	3	4	5	6	7	8	9
1	1	2	3	4	5	6	7	8	9
2	2	4	6	8	10	12	14	16	18
3	3	6	9	12	15	18	21	24	27
4	4	8	12	16	20	24	28	32	36
5	5	10	15	20	25	30	35	40	45
6	6	12	18	24	30	36	42	48	54
7	7	14	21	28	35	42	49	56	63
8	8	16	24	32	40	48	56	64	72
9	9	18	27	36	45	54	63	72	81

Often multiplication tables have ten rows/columns, but since multiplication by 10 is really no different than multiplying by 1 we omit those rows.

Among the 81 entries in this table, there are 18 that begin with a 1; here they are:

$1 \times 1 = 1$ $2 \times 5 = 10$ $2 \times 6 = 12$ $2 \times 7 = 14$ $2 \times 8 = 16$
$2 \times 9 = 18$ $3 \times 4 = 12$ $3 \times 5 = 15$ $3 \times 6 = 18$ $4 \times 3 = 12$
$4 \times 4 = 16$ $5 \times 2 = 10$ $5 \times 3 = 15$ $6 \times 2 = 12$ $6 \times 3 = 18$
$7 \times 2 = 14$ $8 \times 2 = 18$ $9 \times 2 = 18$

On the other hand, there are only three entries that begin with a 9, namely:

$$1 \times 9 = 9 \qquad 3 \times 3 = 9 \qquad 9 \times 1 = 9$$

Here's the tally for leading digits in a standard multiplication table.

First digit	Count	Percentage
1	18	22.22%
2	15	18.52%
3	11	13.58%
4	12	14.81%
5	6	7.41%
6	7	8.64%
7	4	4.94%
8	5	6.17%
9	3	3.70%

We see lower digits are more common than larger digits, but this is not the distribution Benford's law predicts.

A MULTIPLICATION TABLE GIVES ALL POSSIBLE RESULTS of multiplying a single-digit number with another single-digit number. In other words, it gives the result of all pairwise multiplications 1×1, 1×2, and so forth up to 9×9.

To extend this idea, let's consider the results of all possible three-way multiplications of single digits. That is, we do all of the following calculations:

$$1 \times 1 \times 1, \quad 1 \times 1 \times 2, \quad \cdots, \quad 9 \times 9 \times 8, \quad 9 \times 9 \times 9$$

for a grand total of $9^3 = 729$ calculations. After completing these calculations, we tally how often each digit is leftmost and arrive at the following chart:

We can visualize this as a three-dimensional multiplication chart—a $9 \times 9 \times 9$ Rubik's cube with the results (such as $4 \times 7 \times 3 = 84$) inconveniently buried in the interior.

First digit	Count	Percentage
1	218	29.90%
2	137	18.79%
3	94	12.89%
4	81	11.11%
5	46	6.31%
6	43	5.90%
7	37	5.08%
8	37	5.08%
9	36	4.94%

There's no reason to stop at a three-way multiplication. We can create multiplication tables that combine four, five, six, or more numbers at a time. Let's see what happens with a tenfold multiplication table. Such a table would consider all possible multiplications of ten numbers (chosen from 1 through 9). In other words, we calculate all of the following:

A tenfold multiplication table has 9^{10} entries; that works out to be nearly 3.5 billion numbers.

$$1 \times 1 \times 1 \times 1 \times 1 \times 1 \times 1 \times 1 \times 1 \times 1$$
$$1 \times 1 \times 1 \times 1 \times 1 \times 1 \times 1 \times 1 \times 1 \times 2$$
$$1 \times 1 \times 1 \times 1 \times 1 \times 1 \times 1 \times 1 \times 1 \times 3$$
$$\vdots$$
$$1 \times 1 \times 1 \times 1 \times 1 \times 1 \times 1 \times 1 \times 1 \times 9$$
$$1 \times 1 \times 1 \times 1 \times 1 \times 1 \times 1 \times 1 \times 2 \times 1$$
$$1 \times 1 \times 1 \times 1 \times 1 \times 1 \times 1 \times 1 \times 2 \times 2$$
$$1 \times 1 \times 1 \times 1 \times 1 \times 1 \times 1 \times 1 \times 2 \times 3$$
$$\vdots$$
$$9 \times 9 \times 9 \times 9 \times 9 \times 9 \times 9 \times 9 \times 9 \times 8$$
$$9 \times 9 \times 9 \times 9 \times 9 \times 9 \times 9 \times 9 \times 9 \times 9$$

and tally how many of these begin with a 1, how many with a 2, and so on. Here's the result:

First digit	Count	Percentage
1	1048150118	30.06%
2	612266716	17.56%
3	436020803	12.50%
4	342277119	9.82%
5	269493994	7.73%
6	248886318	7.14%
7	188191500	5.40%
8	175747495	5.04%
9	165750338	4.75%

The frequency of the initial digits is a good match with Benford's prediction.

Catching crooks cooking books

BEFORE WE EXPLORE THE DETAILS OF BENFORD'S LAW we briefly mention one practical application. Suppose a dishonest person is filing phony fiscal reports (expense claims, fabricated balance sheets, and so on). In short, the person is lying and is just making up the numbers he or she claims to be real. In a misguided attempt to make the data look realistic, the criminal might inadvertently choose the first digits of the falsified figures evenly from the numerals 1 through 9.

A forensic accountant can quickly check if the initial digits follow Benford's law. If not, that suggests—but does not prove—that the numbers reported are fake.

Refining the problem with scientific notation

SCIENTIFIC NOTATION IS A HANDY WAY to express numbers that are especially large or small. This method expresses numbers such as 12,300,000 as 1.23×10^7. That is, we write a number as a decimal value (that is at least 1 and less than 10) multiplied by a power of 10. We call the decimal value the *mantissa*. For example, the mantissa of 853,100,000 is 8.531:

$$853{,}100{,}000 = \underbrace{8.531}_{\text{mantissa}} \times 10^8.$$

The mantissa of 0.0043 is 4.3 because $0.0043 = 4.3 \times 10^{-3}$.

We've arranged this definition so that the mantissa (of a positive number) is never less than 1 nor is it ever equal to or greater than 10.

$1 \leq \text{mantissa} < 10$

We can use mantissas to present a refinement of Benford's law. Roughly stated, Benford's law asserts that given a large collection of measurements over a wide range of values, about 30% of the numbers will have 1 as their most significant digit. Stated differently, about 30% of the measurements' will have a mantissa m satisfying $m < 2$.

To refine Benford's law, we can look at the first *two* digits of the many measurements and ask: What is the frequency with which the mantissa satisfies (say) $m < 1.7$.

> Here is an equivalent way to ask this question: How often are the first two digits of measurements one of 10, 11, 12, 13, 14, 15, or 16?

More generally, for any number m between 1 and 10, we define $f(m)$ to be the proportion of measurements whose mantissa is less than m.

For example, $f(2)$ is the proportion of measurements with initial digit 1. The value of $f(3)$ gives the proportion with initial digits smaller than 3; that is, measurements with initial digit 1 or 2.

> The notation $f(m)$ is the proportion of measurements whose mantissa is less than m. This bit of notation is important to understanding how the values in Benford's law arise.

How can we use this notation to express the proportion of measurements with initial digit, say, 4? Let's work this out:

- Note that $f(4)$ does not give the proportion with first digit 4; its value is the proportion of measurements with first digit *less than* 4. These are the numbers with first digit 1, 2, and 3.

- Likewise $f(5)$ gives the proportion of measurements with first digit less than 5. These have first digit 1, 2, 3, and 4.

- So, to focus just on the measurements with first digit 4, we subtract $f(5) - f(4)$. The $f(5)$ term includes values with initial digit 1 through 4, and then we take away those values with initial values 1 through 3, leaving only the numbers we're interested in: the ones beginning with a 4.

> To find the proportion of measurements whose first digit is a 4 we calculate $f(5) - f(4)$.

Here are two special values to think about: What are the values of $f(1)$ and $f(10)$? Think about this for a moment before reading on.

REMEMBER THAT $f(m)$ IS THE proportion of measurements whose mantissa is less than m. And remember that mantissas are numbers m that satisfy $1 \leq m < 10$. What are the implications?

- That no number has mantissa less than 1 means that $f(1) = 0$. There are simply no measurements of this form!

- That every number has mantissa less than 10 means that $f(10) = 1$ (or 100% if you prefer). Every number fits the bill.

Between these extremes the value of $f(m)$ increases. The larger m is, the more numbers there are whose mantissas lie below m.

Our next step is to understand how $f(m)$ behaves for values of m between 1 and 10. Our initial concern was with first digits and this notation expands our view to greater generality that will reveal what's happening.

Yards or feet?

WE COLLECT THOUSANDS OF MEASUREMENTS in kilometers and we see a pattern for the initial digits. If we change from kilometers to miles, the same pattern persists. Likewise, if we measure countries' gross domestic products in U.S. dollars we see a pattern in the initial digits of the data. And the same pattern persists when we change from U.S. dollars to Euros (or British pounds or Russian rubles). Let's take a close look at the conversion from yards to feet.

Suppose we measure a great many distances in yards and tally the first digits. How many of these measurements begin with a 2? This tally would include values such as 2.1 or 28 or 0.213 or 299.8 yards. Using the notation we developed in the previous section, the proportion that begins with a 2 is $f(3) - f(2)$.

Now let's convert all the measurements into feet. To do so, we simply multiply by 3. So 2.1 yards be-

For readers whose standard unit of length is a meter, all you need to know is that yard is about the same length as a meter (a bit shorter) and a foot is exactly one-third of a yard.

Remember that $f(3)$ gives the proportion of measurements whose mantissa is less than 3, so that includes all values that start with a 1 or a 2. From that we subtract all those that start with a 1—that's $f(2)$ because that gives the proportion whose initial digit is less than 2.

comes 6.3 feet. Measurements in yards that begin with a 2 result in feet measurements whose first digit ranges from 6 up to, but not including, 9. Surprised?

You might have guessed that if a measurement in yards begins with a 2 then its conversion to feet would begin with a 6. That's not quite right because a measurement such as 2.8 yards is equal to 8.4 feet. So as the mantissa of the yard measurement goes from 2 up to, but not including, 3, the same measurement's mantissa runs from 6 up to, but not including, 9 when expressed in feet.

What proportion of measurements begin with a 6, 7, or 8? The answer is $f(9) - f(6)$.

The $f(9)$ term gives the proportion of measurements whose first digit is 1 through 8, and from that we subtract $f(6)$ to delete those whose first digit is 1 through 5. This leaves the proportion of measurements whose most significant digit is 6, 7, or 8.

Here's the punch line: Since we're dealing with exactly the same collection of values—in yards that begin with a 2 or in feet that begin with 5, 6, or 7—these proportions must be the same: that is to say, $f(3) - f(2)$ gives the same proportion as $f(9) - f(6)$. This is illustrated in the following diagram. Both rectangles represent all of the measurements we recorded; in the left rectangle the measurements are recorded in yards while in the right rectangle the same measurements are in feet. The shaded region on the left represents all those measurements (in yards) that begin with a 2, and the shaded region on the right shows all those measurements (in feet) that begin with a 6, 7, or 8.

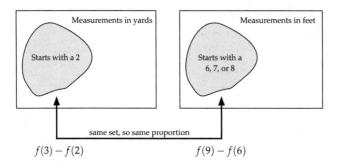

The important thing to notice is that the two shaded regions are identical! So the proportion of measurements beginning with a 2 (on the left) equals the proportion that begins with a 6, 7, or 8 (on the right). Hence the equality $f(3) - f(2) = f(9) - f(6)$.

WE NOW EXTEND THIS IDEA to greater generality. Let's imagine we collect a bunch of measurements, and we count how many of those measurements' mantissas are less than some number, a. The proportion of the measurements that meet this condition is $f(a)$.

Now we switch units, and let's use b to stand for the conversion factor. In other words, if we initially measure an object and the result is 23.5 in the first type of units, then when we redo the measurement in the new set of units the result is $23.5 \times b$.

If we were converting yards to feet, then the conversion factor is $b = 3$. For other conversions, we'd use a different multiplier.

Recall that $f(a)$ gives the proportion of measurements, in the first set of units, with mantissa from 1 up to, but not including, a. The same measurements, when multiplied by b, will have mantissa from a up to, but not including, ab in the second set of units. The proportion of values with mantissa less than a in the original measurements is $f(a)$. The proportion of values (in the new set of units) with mantissa from b up to, but not including ab, is $f(ab) - f(b)$.

And here's the same punch line: The two sets of measurements are the same, so they represent the same proportion of the whole set of values. In symbols we have

There's a "gotcha" if $ab > 10$ or $ab < 1$. Dealing with this restriction is an annoying technicality that we could resolve, but that effort would take us off track and distract us. Furthermore, there's no problem simply continuing our discussion under the assumption that $1 \leq ab \leq 10$.

$$f(a) = f(ab) - f(b),$$

which we rearrange in the following form:

$$f(ab) = f(a) + f(b). \qquad (*)$$

This leads us to the next question: What sort of function satisfies the relation in equation $(*)$ as well as the conditions $f(1) = 0$ and $f(10) = 1$?

What's logs got to do with it?

SOME MATHEMATICAL OPERATIONS CAN BE UNDONE. For example, if we square a positive number we get a result; say, we square 6 to get the value $6^2 = 36$. We undo squaring by taking a square root: $\sqrt{36} = 6$. For positive numbers, the operations of squaring and taking square roots undo each other. In a similar vein, the operation of exponentiating is "undone" by a *logarithm*.

This section presents a refresher about common (base ten) logarithms. If this is familiar to you, skip ahead to the next section.

In this chapter, when we speak of *exponentiation* we mean raising the number 10 to some power. For example, 10^4 yields the result 10,000. We "undo" this raising to a power via the logarithm function:

$$\log(10{,}000) = 4.$$

A useful way to interpret the log function is to think of it as the "What exponent?" function. What exponent do we place on the number 10 to yield a prescribed value? For example, what exponent would give us 1000? Since $1000 = 10 \times 10 \times 10 = 10^3$, the exponent that yields 1000 is 3. That's precisely what we mean when we write $\log(1000) = 3$.

It's not difficult to understand what happens when we raise 10 to a positive integer exponent; we simply multiply 10 by itself that many times:

$$10^6 = \underbrace{10 \times 10 \times 10 \times 10 \times 10 \times 10}_{\text{six terms}} = 1{,}000{,}000.$$

And so, conversely, finding the logarithm of a power of 10 is a simple matter of counting zeros:

$$\log(1{,}000{,}000{,}000) = 9.$$

Raising 10 to a non-integer power is more complicated. The key idea is to understand the result of multiplying 10^m and 10^n for exponents m and n.

What is the result of multiplying $10^6 \times 10^5$? Fear not, because multiplying by 10 is easy. To see what's going on, let's write out exactly what this means:

$$10^6 \times 10^5 = \left(\underbrace{10 \times 10 \times 10 \times 10 \times 10 \times 10}_{\text{six terms}}\right) \times \left(\underbrace{10 \times 10 \times 10 \times 10 \times 10}_{\text{five terms}}\right).$$

What's the result? No need to multiply! Just count the factors of 10: there are eleven of them. In other words:

$$10^6 \times 10^5 = 10^{11}.$$

Once we see this, we realize that for positive integer exponents, we have

$$10^m \times 10^n = 10^{m+n} \qquad \text{(law of exponents)}$$

because 10^m contributes m factors of 10, and 10^n contributes an additional n factors of 10.

The central idea to extending exponentiation to non-integer powers is to extend the law of exponents $10^m \times 10^n = 10^{m+n}$ for all possible exponents. Let's see where that takes us.

Let's work out $10^{0.5}$. We might not know what that number is, but let's see what happens when we multiply $10^{0.5} \times 10^{0.5}$. Setting both $a = m$ and n equal to 0.5 in the relation $10^m \times 10^b n = 10^{m+n}$ we have this:

$$10^{0.5} \times 10^{0.5} = 10^{0.5 + 0.5} = 10^1 = 10.$$

What we've learned is that if we multiply $10^{0.5}$ by itself, the result is 10. In other words, $10^{0.5}$ is the square root of 10:

$$10^{0.5} = \sqrt{10} \approx 3.162.$$

We encourage you to try this on a calculator. Compute both $10^{0.5}$ and $\sqrt{10}$ and observe that both give the same result.

With more work (that would take us much too far afield) we can calculate all powers of 10. The following figure plots the graph of the function 10^x as x ranges from 0 to 1.

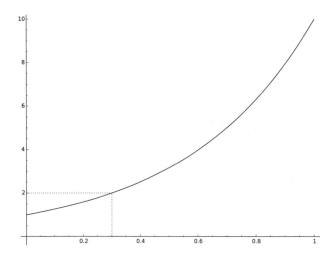

For what value of x does $10^x = 2$? Looking at the graph of 10^x suggests that $x = 0.3$ fits the bill. Try this on your calculator and you'll find that $10^{0.3} = 1.99526\ldots$, which is close, but not exactly equal to 2. We need to increase the exponent just a little; let's try $10^{0.301}$; the result is 1.99986. Closer, but not quite right. We need something a little bigger. The value x that yields $10^x = 2$ is $0.30102999566398114\ldots$. And it is that value that is precisely $\log(2)$. (You've seen this number earlier in this chapter. Go find it!)

THE EXPONENTIATION LAW $10^m \times 10^n = 10^{m+n}$ can be re-expressed in terms of logarithms. To see how, let $a = 10^m$ and $b = 10^n$.

What is the logarithm of a? That's the exponent we need to place on 10 to get the value a. Since $a = 10^m$, that means $\log(a) = m$. Likewise, $\log(b) = m$.

What is the logarithm of $a \times b$? We know that $a = 10^m$ and $b = 10^n$ and so $a \times b = 10^{m+n}$. What exponent do we need to place on 10 to get the value $a \times b$? The answer is $m + n$. In symbols: $\log(a \times b) = m + n$.

Recapping, we have the following:

$$\log(a) = m, \quad \log(b) = n, \quad \text{and} \quad \log(a \times b) = m + n.$$

Combining these together yields the following *law of logarithms*:

$$\log(ab) = \log(a) + \log(b). \qquad (**)$$

Now where have we seen this sort of relation before?

Collecting the loose ends

LET'S TIE ALL THIS TOGETHER. We defined a function f that gives the proportion of measurements whose mantissa is less than some value m. It satisfies these three properties:

$$f(1) = 0, \quad f(10) = 1, \quad \text{and} \quad f(ab) = f(a) + f(b).$$

Then we paused to discuss logarithms and found that the log function satisfies these properties:

$$\log(1) = 0, \quad \log(10) = 1, \quad \text{and} \quad \log(ab) = \log(a) + \log(b).$$

In other words, both f and log have the same value at 1 and at 10, and they obey the same relation as expressed in equations $(*)$ and $(**)$. With this (and the technical proviso that f is continuous) mathematicians can prove that f must be exactly the same as log, and now we're ready to work out the distribution of initial digits of measurements.

What proportion of measurements begin with a 1? That's the same as asking what proportion have a mantissa that is less than 2. So the answer is $f(2)$, which we now know equals $\log(2) \approx 0.3010 = 30.1\%$.

What proportion of measurements begin with a 9? The answer is $f(10) - f(9)$ (subtract from all measurements, $f(10)$, those measurements whose mantissa is less than 9). We have

$$f(10) - f(9) = \log(10) - \log(9) \approx 1 - 0.9542 = 0.0458 = 4.58\%.$$

On page 116 we asked for the value of $f(1.7)$ and now we can answer. We have $f(1.7) = \log(1.7) \approx 0.23 = 23\%$.

12
Algorithm

CREATIVE CHEFS GENERALLY DON'T FOLLOW RECIPES EXACTLY. Rather, they use the instructions to inspire their cooking. Novice cooks are more likely to follow the steps slavishly.

Similarly, drivers with an excellent sense of direction don't need maps or written instructions to find their destinations. Others want detailed turn-by-turn guidance.

Computers are like the newbies. When given a list of numbers to add, they follow a meticulously prescribed set of steps, performing each operation according to a programmed procedure. These procedures are known as *algorithms*. Computer algorithms are omnipresent in our lives: they add interest to our bank accounts, determine the location of page breaks in a text document, convert digital data on DVDs into movies, predict the weather, search the web for recipes containing a given list of ingredients, and speak to us from our GPS devices as we try to find an obscure address.

The first mathematical algorithm most people learn is how to add numbers. Asked to add $25 + 18$, we know that we first add the 5 and the 8 (we've memorized the result: 13), write down the 3, carry the 1, and so on.

An algorithm designer does more than work out a

> The word *algorithm* derives from the name of a ninth-century Persian mathematician named Al-Khwarizmi.

correct procedure for solving a problem; the method should also be efficient. An algorithm that is mathematically correct but takes centuries to complete its work is of little use. Let's look at come examples.

Sorting

AT THE END OF EVERY SEMESTER I HAVE A STACK of final exam papers to return to my students. When students come to my office to collect their work, I don't want to rummage through a disorganized pile to find their test. Rather, I keep the papers arranged alphabetically by student name. So, before I announce that the tests are ready for pickup, they need to be sorted.

The problem is to take a pile of papers in some unknown order and rearrange them into a new pile that is in alphabetical order. What's the best way to do this?

LET'S BEGIN WITH A SIMPLE BUT RATHER INEFFICIENT IDEA. Suppose I have one hundred students in my class. I take the first test paper from the unsorted pile and I see if it happens to be first alphabetically. How do I do this? I compare this paper against all the other papers in the pile. Chances are, this paper on the top of the unsorted pile is not the first alphabetically, so I put it at the bottom of the pile and try again. I keep doing this until I identify the paper that's first alphabetically. I remove that paper and place it into a new, second pile that will be used for collecting the papers alphabetically.

Now I return to the unsorted pile—it has 99 papers now—and, just as before, look for the paper that comes first alphabetically. I do this by taking the top paper in the pile, comparing it against all the other papers in the pile, and placing it at the end if it's not the right one. When I find the one that's alphabetically smallest, I remove it from the unsorted pile and place it at the end of the sorted pile.

The unsorted pile now has "only" 98 papers left, and I repeat the procedure: hunt for the paper that is smallest alphabetically and then move it over to the end of the sorted pile.

How long does this take?

The fundamental step is to look at two papers and decide which comes first alphabetically. We assess the efficiency of a sorting procedure by counting the number of these basic comparisons it performs. Since my class has one hundred students, how many times do I have to hold up a pair of exams, read the names, and decide which comes first?

> Even this fundamental step can be broken down into more basic steps. For example, to decide which comes first, ALICE or ALEX, we start by comparing the first letters. It's A in both cases, so we compare second letters. Again, it's a tie (both are L) so we go on to the third letter. Since E comes before I, we conclude that ALEX precedes ALICE.

In the unordered stack of 100 papers, I compare the first paper with each of the next: that's 99 comparisons. I might have to do this for all 100 papers (the paper I'm looking for might be the last one in the pile). So to find the first paper alphabetically might take $100 \times 99 = 9900$ comparisons.

Having removed that paper and placed it on the sorted pile, I repeat the procedure on the stack of 99 unsorted papers. I compare the first paper to the 98 others to see if it's the first alphabetically, and I may have to do this analysis for all the papers in the unsorted stack. Finding this second paper might take $99 \times 98 = 9702$ comparisons.

Finding the third paper takes 98×97, the fourth takes 97×96, and so forth. To sort the entire stack requires a whopping

$$100 \times 99 + 99 \times 98 + 98 \times 97 + \cdots + 2 \times 1 = 333{,}300$$

comparisons.

We have performed a *worst case* analysis. For each calculation we presumed the worst and calculate how many comparisons we'd have to make. While worst case analysis is undoubtedly overly pessimistic, it does give us a sense of just how inefficient this

> An alternative to worst case analysis is *average case analysis* in which we compute the number of comparisons in a typical situation.

method is. Let's try another.

WE START WITH A STACK of 100 papers in some mixed-up order. We begin by looking at the first two papers in the stack. If they are in the wrong order, we swap their positions (first becomes second and second becomes first). If they're in the right order, we leave them alone. Now we look at the second and third papers in the stack. If they're in the right order we leave them alone, but if they're in the wrong order we swap them. And we continue this way through the entire stack. One such pass requires 99 comparisons.

After this pass through the stack, papers early in the alphabet tend to move toward the top of the pile and papers late in the alphabet sink to the bottom. But one pass through the stack does not necessarily place the papers in order. In the worst case, the paper that is first alphabetically may begin at the bottom of the stack. One pass through the pile only advances it to the 99th position. It will take 99 iterations of the procedure to be sure we have sorted the pile.

Therefore, the number of comparison made by this method is $99 \times 99 = 9801$. This is much better than the first method but still onerous. If I can compare two papers and (if need be) swap them in just two seconds, it will take me over five hours to get these papers in alphabetical order. This is intolerable.

AT THIS POINT I'M DEPRESSED, SO I LEAVE MY OFFICE FOR A WALK. Down the hall, I see the two postdoctoral fellows who work for me and an evil grin comes to my lips. Quickly, I run back to my office, divide the unsorted pile of exams in half, and give fifty to each of my postdocs. "Here's a pile of papers for each of you," I say. "Please arrange each of your stacks in alphabetical order and return them to my office when you're done." Happy, I retreat to

The procedure we describe here is known as *bubble sort*. The following diagram shows one pass of the algorithm.

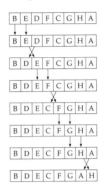

Notice that A has moved only one position toward the front of the pile. It will take six more rounds for it to bubble up to its proper location.

The algorithm we describe here is known as *merge sort*. It is an example of a *divide-and-conquer* strategy for problem solving: we take a large task, break it into smaller, more manageable pieces, solve those subproblems, and then combine the answers to the subproblems.

my office.

I still will have some work to do when my postdocs have finished sorting their piles. I will need to merge their separate stacks into a finished pile. How hard will that be? I'll have these two sorted piles sitting on my desk. I will look at the top paper on each of the piles and whichever paper is closer to the front of the alphabet I place on the final combined pile. This diagram illustrates this merging process:

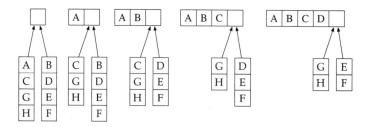

When one of the piles runs out, I just place the papers in the other pile at the end of the sorted pile. In the worst case, I've made only 99 comparisons. I can do that in just a few minutes!

But what about my postdocs? Each has a stack of 50 papers to sort. The post docs are extremely bright, so instead of sorting their piles themselves, they both divide their piles in half (so there are four unsorted piles of 25 papers each) and ask four graduate students to sort their piles! When the grad students are finished with their work, the postdocs just have to merge two sorted piles of 25 into single sorted piles of 50. Each postdoc makes (at most) 49 comparisons.

Now the four graduate students are no fools. They split their piles in two (12 and 13 papers each) and find (collectively) eight senior undergraduates to sort the small piles. The grad students will still need to merge the two piles they each get back from the seniors before handing their piles of 25 to the postdocs.

How do the seniors sort their piles? You guessed

it: They break their piles in half (with 6 or 7 in each) and find juniors to sort the smaller piles. The juniors, in turn, split the piles in half (with 3 or 4 in each) and hand them off to sophomores. Finally, the sophomores split their piles in two (piles now of size 1 or 2) and give them to a hoard of freshmen. The freshmen, with no place to turn, simply sort their piles—this isn't difficult because their piles have either 1 or 2 papers!

It's pretty clear that this Ponzi scheme saves *me* time, but what of the total effort? Let's do an accounting of how many comparisons all the people in this enterprise make. Here's a chart that accounts for all the work:

Person Sorting	Number of Workers	Comparisons Per Worker	Total Number of Comparisons
Me	1	99	99
Postdocs	2	49	98
Grad Students	4	24	96
Seniors	8	12	96
Juniors	16	6	96
Sophomores	32	3	96
Freshmen	64	1	64
		Grand total	645

The total effort in this method is much less than the bubble sort method we described earlier.

UNFORTUNATELY, THERE'S ONLY ONE TINY FLAW in this idea: I don't have any postdoctoral fellows working for me! So instead of creating a veritable army of assistants to get my exam papers in alphabetical order, I'm going to do this process myself.

I start by finding a large, empty conference table. I take my stack of 100 papers and divide it into two piles of 50 each, placing these piles at opposite ends of the table. I separately sort the two piles and then merge the two piles together. How do I sort the piles? I use exactly the same algorithm! I break each pile of 50 into two piles of 25, sort, and then merge. Each pile of 25 is sorted by the same sort-merge method. The total number of comparisons made is the same as before (645) only now I'm doing all the work. Still this is much less work than the nearly ten thousand comparisons bubble sort takes.

A DICTIONARY DEFINITION OF A WORD should not include the word being defined. Imagine looking up the word *poverty* in a dictionary only to find this definition:

Poverty: Being in a state of poverty.

Worthless! Interestingly, however, the definition of the merge-sort algorithm commits this self-referential "sin." Here's a more formal description of merge sort (with all the details omitted).

MERGE-SORT ALGORITHM

Input: A list of items $a_1, a_2, \ldots, a_n$.
Output: The items listed in sorted order.

1. If $n = 1$, the list is already in sorted order; just return it as output. Otherwise, continue to step 2.

2. Split the input list into two (nearly) equal sublists. Use MERGE-SORT to sort the sublists.

3. Merge the sublists to create the output list.

This description of the merge-sort algorithm would be complete if we included an explanation of how the merge step works.

An algorithm that is defined in terms of itself is called *recursive*. As opposed to our impoverished definition of *poverty*, this definition is acceptable because the self-referencing loop eventually ends. Thanks to step 1, once the list is short enough, the merge-sort procedure does not revisit itself; there is no "infinite descent."

Greatest common divisor

OF THE NUMBERS THAT DIVIDE INTO BOTH 986 AND 748 EXACTLY, which is the largest? The simplest way to approach this problem is to start trying possibilities. Obviously both 986 and 748 are divisible by 1. And it's easy to see that both are divisible by 2. Neither is divisible by 3. One of them, 748, is divisible by 4 but the other is not. "All" we need to do is to keep trying possible divisors and keep track of the results. We stop testing divisors when we reach 748 because no integer larger than that can be a factor of 748. After trying all the possibilities, we find that the only common divisors of 986 and 748 are 1, 2, 17, and 34. The *greatest common divisor* of 986 and 748 is 34. For any two positive integers a and b, the notation $\gcd(a,b)$ stands for their greatest common divisor.

Some easy examples for you to check your understanding:

$\gcd(10,15) = 5$
$\gcd(12,16) = 4$
$\gcd(13,11) = 1$
$\gcd(10,20) = 10$
$\gcd(17,17) = 17$

The method we described above yields a simple, correct algorithm for calculating the greatest common divisor of two positive integers. Its weakness is that it is inefficient. To find the gcd of two three-digit numbers requires us to try hundreds of possible divisors. Is there a better way?

LET'S LOOK AT THE NUMBERS 986 AND 748 MORE CLOSELY. Since we are looking for common factors, it seems natural to factor them into primes (see Chapter 1). Here are their prime factorizations:

$$986 = 2 \times 17 \times 29 \quad \text{and} \quad 748 = 2 \times 2 \times 11 \times 17.$$

With these factorizations in hand, we can work out the gcd by "grabbing" all the primes the two numbers have in common. Since they both have one factor of 2 and one factor of 17, their gcd is $2 \times 17 = 34$.

How do we factor numbers efficiently? The unfortunate answer (which we mentioned in Chapter 1) is

When calculating $\gcd(a,b)$, factoring a and b is more efficient than trial division up to the lesser of a and b. To find the prime factors of an integer a requires at most $\sqrt{a}$ divisions. So this is a significant improvement over the first approach, but still infeasible for large (100-digit) integers.

that we don't know how. We need a new idea.

THE NEW IDEA COMES FROM EUCLID. Suppose d is a common divisor of 986 and of 748. This means

$$986 = xd \quad \text{and} \quad 748 = yd$$

where x and y are integers. This implies that d is a divisor of $986 - 748$. Here's the algebra to back up this claim:

$$986 - 748 = xd - yd = (x - y)d.$$

Since x and y are integers, so is $x - y$. Therefore the difference between 986 and 748 is also a multiple of d. Note that $986 - 748 = 238$.

Likewise, any common divisor of 748 and 238 is a divisor of 986; here's why. If e is a common divisor of 748 and 238 we have that

$$748 = ue \quad \text{and} \quad 238 = ve$$

where u and v are integers. Therefore

$$986 = 748 + 238 = ue + ve = (u + v)e,$$

from which we deduce that 986 is a multiple of e.

Conclusion: The common divisors of 986 and 748 are exactly the same as the common divisors of 748 and 238. To see this explicitly, here are the divisors for all three numbers with the divisors they share underlined:

divisors of 986 ⟶ $\underline{1}, \underline{2}, \underline{17}, 29, \underline{34}, 58, 493, 986$
divisors of 748 ⟶ $\underline{1}, \underline{2}, 4, 11, \underline{17}, 22, \underline{34}, 44, 68, 187, 374, 748$
divisors of 238 ⟶ $\underline{1}, \underline{2}, 7, 14, \underline{17}, \underline{34}, 119, 238$.

Because their common divisors are the same, we have that

$$\gcd(986, 748) = \gcd(748, 238). \quad \text{(A)}$$

Therefore the problem of computing gcd(986, 748) is replaced by gcd(748, 238). This is progress because we are calculating with smaller numbers. Let's try the same trick again.

If d is a common divisor of 748 and 238, then it's also a divisor of their difference. But we can say even more. We can subtract 238 several times from 748 and make the same claim. To be specific, if d divides 748 and 238, then we claim that d is also a factor of $748 - 3 \times 238$. Here's the algebra to support this. Write
$$748 = xd \quad \text{and} \quad 238 = yd$$
where x and y are integers. Then
$$748 - 3 \times 238 = xd - 3yd = (x - 3y)d$$
and hence d is a factor of $748 - 3 \times 238 = 34$. Conversely, if e is a divisor of 34 and 238 then e is also a divisor of 748. Here's the algebra. Since e is a factor of 238 and 34 we have
$$238 = ue \quad \text{and} \quad 34 = ve$$
where u and v are integers. Therefore
$$748 = 3 \times 238 + 34 = 3 \times (ue) + ve = (3u + v)e$$
and so e is a divisor of 748. Hence the three integers 748, 238, and 34 have precisely the same divisors and we may conclude that
$$\gcd(748, 238) = \gcd(238, 34). \quad \text{(B)}$$
Combining equations (A) and (B) we have
$$\gcd(986, 748) = \gcd(748, 238) = \gcd(238, 34).$$

We're nearly done. Notice that 238 is divisible by 34 (because $238 = 34 \times 7$) and so $\gcd(238, 34) = 34$. The grand finale is
$$\gcd(986, 748) = \gcd(748, 238) = \gcd(238, 34) = 34.$$

Let's review how we came up with the intermediate numbers in this calculation. We subtracted 748 from 986 to get 238. We subtracted 238 three times from 748 to get 34. Why did we subtract once in the first instance but thrice in the second? We want to reduce the numbers as much as possible because computing gcd's of smaller numbers is easier than for larger. So we want to subtract the lesser number from the larger as many times as possible. Notice that 748 goes into 986 only once (in the sense of division) leaving a remainder of 238. However, 238 goes into 748 three times, leaving a remainder of 34. We can only subtract 748 from 948 once, but we can subtract 238 from 748 three times.

WE ARE READY TO ASSEMBLE THE IDEAS from this example into an algorithm for computing the greatest common divisor of two positive integers. We are given two positive integers a and b and we want to compute their greatest common divisor $\gcd(a, b)$. We are going to subtract as many multiples of b from a as possible. To figure out how many, we divide a by b and get the quotient q and remainder c. Algebraically, we have

$$a - qb = c.$$

If b happens to be a divisor of a, then the division is exact and the remainder c is zero. Otherwise, c is positive, but less than b (if c were larger than b, we could subtract another copy of b).

Now suppose d is a common divisor of a and b. This means

$$a = xd \quad \text{and} \quad b = yd$$

where x and y are integers. Therefore

$$c = a - qb = xd - q(yd) = (x - qy)d$$

and so c is also a multiple of d (because $x - qy$ is an integer).

For example, if $a = 100$ and $b = 40$, then the quotient $q = 2$ and the remainder $c = 20$. This gives

$$100 - 2 \times 40 = 20.$$

On the other hand, if e is a common divisor of b and c, then we have that

$$b = ue \quad \text{and} \quad c = ve$$

where u and v are integers. Therefore

$$a = c + qb = ve + q(ue) = (v + qu)e$$

and so e is a divisor of a.

What we have proved is that the common divisors of a and b are exactly the same as the common divisors of b and c. Ergo,

$$\gcd(a,b) = \gcd(b,c). \tag{C}$$

Let's see how equation (C) enables us to efficiently calculate the greatest common divisor of two large integers: $a = 10693$ and $b = 2220$.

We divide a by b and find that 2220 goes into 10693 four times and leaves a remainder of $c = 1813$. Therefore

$$10693 = 4 \times 2220 + 1813.$$

$$\gcd(10693, 2220) = \gcd(2220, 1813).$$

Now we "start over" letting $a' = 2220$ and $b' = 1813$. Dividing a' by b' we see that 1813 goes into 2220 only once and leaves a remainder $c' = 407$. Therefore, by equation (C):

$$2220 = 1 \times 1813 + 407.$$

$$\gcd(10693, 2220) = \gcd(2220, 1813) = \gcd(1813, 407).$$

Starting over again, this time with $a'' = 1813$ and $b'' = 407$ we find that 407 goes into 1813 four times leaving a remainder of 185. Again, by equation (C):

$$1813 = 4 \times 407 + 185.$$

$$\gcd(10693, 2220) = \gcd(2220, 1813) = \gcd(1813, 407) = \gcd(407, 185).$$

Put $a''' = 407$ and $b''' = 185$. Dividing, we see that 185 goes into 407 twice leaving a remainder of $c''' = 37$.

$$407 = 2 \times 185 + 37.$$

$$\gcd(10693, 2220) = \gcd(2220, 1813) = \gcd(1813, 407)$$
$$= \gcd(407, 185) = \gcd(185, 37)$$

We're getting close! Put $a'''' = 185$ and $b'''' = 37$, divide and—behold!—37 goes into 185 exactly five times. Therefore $\gcd(185, 37) = 37$. We complete our calculation:

$185 = 5 \times 37 + 0.$

$$\begin{aligned}\gcd(10693, 2220) &= \gcd(2220, 1813) \\ &= \gcd(1813, 407) \\ &= \gcd(407, 185) \\ &= \gcd(185, 37) = 37.\end{aligned}$$

We have found the greatest common divisor of 10693 and 2220 with only five divisions!

Euclid's algorithm for the greatest common divisor can be summarized like this:

In Chapter 6 we introduced the concept of integers being *relatively prime*. Here is an alternative description: a and b are relatively prime provided $\gcd(a, b) = 1$. Since Euclid's algorithm efficiently calculates the gcd of two numbers, it provides an effective way to see if two numbers are relatively prime.

Euclid's GCD algorithm

Input: Two positive integers a and b.
Output: $\gcd(a, b)$

1. Divide a by b to find a quotient q and a remainder c.

2. If $c = 0$, then return b; it's the greatest common divisor of a and b.

3. Otherwise, calculate $\gcd(b, c)$ and return that value.

Check your understanding of Euclid's algorithm to find the greatest common divisor of 1309 and 1105. You may use a calculator. Check your result on page 139.

Least common multiple

THE GREATEST COMMON DIVISOR CONCEPT IS CLOSELY RELATED to the notion of *least common multiple*. Given two positive integers, say 10 and 15, their least common multiple is the smallest positive integer that is a multiple of both; in this case the answer is 30. The notation we use for the least common multiple of a and b is $\text{lcm}(a,b)$.

The least common multiple concept is useful when adding fractions. For example, to add $\frac{1}{10}$ and $\frac{1}{15}$ the first step is to rewrite these fractions with a common denominator. That common denominator may be any number that is a multiple of both 10 and 15, the simplest of which is their lcm. Since $\text{lcm}(10,15) = 30$, we can re-express $\frac{1}{10}$ and $\frac{1}{15}$ as fractions with the denominator of 30 and then add:

$$\frac{1}{10} + \frac{1}{15} = \frac{3}{30} + \frac{2}{30} = \frac{5}{30} = \frac{1}{6}.$$

Finding the least common multiple of small numbers is not too challenging, but how do we calculate this for larger numbers? For example, what is the least common multiple of 364 and 286?

One way is to write out multiples of the two numbers and hope we find a match:

multiples of $364 \to 364, 728, 1092, 1456, 1820, 2184, \ldots$
multiples of $286 \to 286, 572, 858, 1144, 1430, 1716, 2002, \ldots$

If we extend these lists far enough, we find that 4004 is the first number they have in common, so $\text{lcm}(364, 286) = 4004$.

This method is inefficient, but it is not hopeless. We know that 364×286 is a multiple of both numbers. We just hope we stumble on a common multiple that's smaller.

HERE'S ANOTHER IDEA. Factor 364 and 286 into primes:

$364 = 2 \times 2 \times 7 \times 13$ and $286 = 2 \times 11 \times 13$.

A multiple of 364 must include the prime factors $2 \times 2 \times 7 \times 13$ and a multiple of 286 must include the prime factors $2 \times 11 \times 13$. So in building a common multiple of these numbers we must have these factors: two 2s, a 7, an 11, and a 13. (We don't need both

13s.) This gives

$$2 \times 2 \times 7 \times 11 \times 13 = 4004$$

and, indeed, 4004 is the least common multiple of 364 and 286.

This seems like a terrific method except—as we explained earlier in this chapter and in Chapter 1—there is no known efficient method for factoring large integers.

Although factoring does not give us an efficient algorithm for computing the lcm of two numbers, it does give us an important insight. Let's compare the factoring methods for gcd and lcm.

The prime seven factors of the two numbers taken together are:

$$\underbrace{2, 2, 7, 13}_{364}, \underbrace{2, 11, 13}_{286}.$$

We form the gcd of 364 and 286 by collecting the prime factors these lists have in common. There are two: 2 and 13.

To create a common multiple of 364 and 286, we need to include all the primes that appear in both lists. However, we don't need two factors of 13 (one is enough) and we don't need three factors of 2 (two are enough). In other words, we form the lcm by taking all the primes on one list and all the primes on the other, but omitting "redundant" primes (primes that are on both lists). There are five: 2, 2, 7, 11, and 13.

Checking:

$$\gcd(364, 286) = 26 = 2 \times 13,$$

and $\text{lcm}(364, 286) = 4004 = 2 \times 2 \times 7 \times 11 \times 13.$

Notice that the primes we used for the lcm include all of the prime factors for the two numbers except the ones we used for the gcd:

$$\underbrace{2 \times 2 \times 7 \times 11 \times 13}_{\text{lcm}} \times \underbrace{2 \times 13}_{\text{gcd}}.$$

In other words, we have

$$364 \times 286 = (2 \times 2 \times 7 \times 13) \times (2 \times 11 \times 13)$$
$$= (2 \times 2 \times 7 \times 11 \times 13) \times (2 \times 13)$$
$$= \text{lcm}(364, 286) \times \gcd(364, 286).$$

There is nothing special about this example. For any two positive integers a and b, we have

$$a \times b = \text{lcm}(a, b) \times \gcd(a, b),$$

which we can rearrange to give the formula

$$\text{lcm}(a, b) = \frac{ab}{\gcd(a, b)}. \tag{D}$$

Since Euclid's algorithm gives an efficient way to calculate the greatest common divisor of two integers, it also gives—thanks to equation (D)—an efficient way to find their least common multiple.

Solution to the GCD problem on page 136. Using Euclid's algorithm:

$$1309 = 1 \times 1105 + 204$$
$$1105 = 5 \times 204 + 85$$
$$204 = 2 \times 85 + 34$$
$$85 = 2 \times 34 + 17$$
$$34 = 2 \times 17 + 0$$

Therefore

$$\gcd(1309, 1105) = \gcd(1105, 204) = \gcd(204, 85)$$
$$= \gcd(85, 34) = \gcd(34, 17) = 17.$$

PART II: SHAPE

13
Triangles

A TRIANGLE IS A GEOMETRIC FIGURE comprised of three line segments connecting three points. In this chapter we'll explore some well-known features of these humble shapes and uncover a few of their mysteries. Let's start with two familiar formulas for triangles: the sum of the angles and the area.

It all adds up to 180

PERHAPS THE MOST FAMILIAR FACT ABOUT TRIANGLES is that when we measure the three angles of a triangle we find that those numbers add up to $180°$.

How do we know that is true? It's not by cutting lots of triangles out of paper and checking their angles with a protractor! Let's see why this works.

Pick a triangle—any triangle—and names its three corners A, B, and C. And let's say that the angles at these corners are x, y, and z degrees respectively. We want to convince ourselves that $x + y + z = 180$.

Draw (either in your mind or on paper) a line L through the point B that is parallel to the line segment AB, like this:

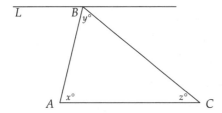

Next we extend line segments AB and BC so they poke through to the other side of line L. Doing so creates three angles on the other side of L.

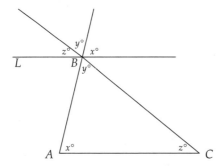

Notice that the three new angles we created, when taken together, exactly fill one side of the line L. This shows that those three angles add up to 180 degrees.

In the diagram we labeled the newly formed angles $x°$, $y°$, and $z°$ because they have the same sizes as the angles in the triangle. Why?

When two parallel lines are cut by another line, the oblique line slices off two equal angles. And when two lines meet at a point, the opposite angles are the same. This is illustrated by the figures in the margin.

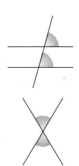

Now look back at the three new angles. Because lines AC and L are parallel, the line AB cuts off equal angles—both are x degrees. Likewise, the line BC cuts off two equal angles—both are z degrees. Finally, the two lines AB and BC meet at the point B and the two angles they form are equal—both are y degrees.

Let's summarize what we know:

- The three new angles span exactly one side of the line L so they add up to 180°.

- The three new angles have the same sizes as the three angles of the triangle.

And so we conclude that $x + y + z = 180$ as promised.

Area

COUNTLESS STUDENTS MEMORIZE that the area of a triangle equals one-half the base times the height. Recall that "the base" is one of its sides and "the height" is the distance from the third corner to the base.

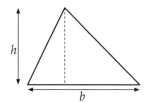

If the length of the base is b and the height is h, then the area of the triangle is $\frac{1}{2} \times b \times h$.

Familiar stuff! But why is it true? There's a lovely explanation that's more interesting than the formula itself.

Make a duplicate copy of the triangle whose area we seek, flip it over, and line it up with the original triangle to make a parallelogram like this:

We show how to derive the formula area = $\frac{1}{2}bh$ from the fact that the area of an a-by-b rectangle is $a \times b$.

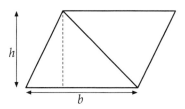

Because this parallelogram is made up of two identical triangles, its area is double the area of the original triangle.

Now we "square up" the parallelogram by lopping off the little triangle sticking out on one side—cut along the dotted line. Take that little triangle and attach it to the opposite side of the parallelogram like this:

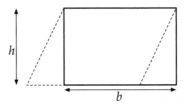

The result is a *b*-by-*h* rectangle whose area is $b \times h$. Since the ingredients of this rectangle were two identical copies of the original triangle, that triangle's area is $\frac{1}{2} \times b \times h$.

IF WE ARE GIVEN A PHYSICAL TRIANGLE, say, one made of wood, it's pretty easy to use a tape measure to find the length of one of its sides. But it's not as convenient to measure the height. We can place the tape measure at one corner but then have to estimate to where on the opposite side we should measure.

So let's suppose we know the lengths of the three sides of a triangle. How can we calculate its area? Would it take a Herculean effort? No, just the help of a Hero—Hero of Alexandria who lived about 2000 years ago.

Suppose that the lengths of the sides of a triangle are *a*, *b*, and *c* as shown in the figure.

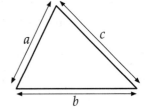

To calculate its area, Hero tells us to first add the lengths of the sides and divide by two. Call the result *s*:

$$s = \frac{1}{2}(a+b+c).$$

The next step is to substitute the four numbers *a*, *b*, *c*, and *s* into this formula to find the area:

$$\text{area} = \sqrt{s(s-a)(s-b)(s-c)}.$$

For example, if the lengths of the sides of a triangle are 4, 5, and 7, then $s = \frac{1}{2}(4+5+7) = \frac{1}{2} \times 16 = 8$. This gives

$$\text{area} = \sqrt{8(8-4)(8-5)(8-7)} = \sqrt{8 \times 4 \times 3 \times 1} = \sqrt{96} \approx 9.8.$$

Here's an alternative version of Hero's formula in which we do not need to first calculate the value s:

$$\text{area} = \frac{1}{4}\sqrt{(a+b+c)(a+b-c)(a+c-b)(b+c-a)}.$$

Returning to the triangle whose sides have lengths 4, 5, and 7, we calculate like this:

$$\text{area} = \frac{1}{4}\sqrt{(4+5+7) \times (4+5-7) \times (4+7-5) \times (5+7-4)}$$
$$= \frac{1}{4}\sqrt{16 \times 2 \times 6 \times 8} = \frac{1}{4}\sqrt{1536} \approx 9.8.$$

THERE ARE SEVERAL MORE FORMULAS for the area of a triangle, but let's round out this section with a personal favorite. This formula applies in the case that the triangle is drawn on graph paper with the three corners lying at grid points as shown in this figure.

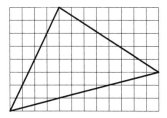

Let's assume the squares in the grid are size 1×1. We might find the area of the triangle by counting the number of grid cells completely contained in the interior of the triangle and then somehow use the boxes that are cut by the sides of the triangle to get the rest of the area. That would be messy.

A cleaner approach is provided by Pick's Theorem. Rather than counting the little squares, we count

Here's another way to find the area of this triangle. Notice that the triangle fits snuggly in an 8-by-12 rectangle whose area is therefore $8 \times 12 = 96$. There are three "scrap" right triangles to throw away. The areas of these disposable triangles are easy to figure out because they are right triangles.

The scrap triangle on the left has side lengths 8 and 4 so its area is $\frac{1}{2} \times 8 \times 4 = 16$. The one on the upper right has side lengths 8 and 5 so its area is $\frac{1}{2} \times 8 \times 5 = 20$. And the one below has side lengths 12 and 3 so its area is $\frac{1}{2} \times 12 \times 3 = 18$.

The total area of the three scrap triangles is $18 + 20 + 16 = 54$. We subtract that from the area of the bounding rectangle to get $96 - 54$, which equals 42.

grid points. First we count the number of grid points lying in the interior of the triangle; call that number I. Then we count the number of grid points on the boundary of the triangle; call that number B.

Then Pick's Theorem tells us that

$$\text{area} = I + \frac{1}{2}B - 1.$$

We drew the triangle in the diagram large so you can count the points. When you do, you should find that there are $I = 38$ grid points contained in the interior of the triangle and $B = 10$ grid points (including the corners) on the boundary. Therefore, by Pick's Theorem we have

$$\text{area} = 38 + \frac{1}{2} \times 10 - 1 = 42.$$

We close this section with a puzzle for you. Suppose we wish to find the area of a quadrilateral drawn on graph paper. Let's assume that the four corners of the quadrilateral lie at grid points. If there are I grid points contained in the interior of the quadrilateral and B on the boundary (including the four corners), what's the area of the quadrilateral? The solution is on page 152.

Now think about the same question for pentagons, hexagons, and so forth.

Centers

WHAT DO WE MEAN BY THE "CENTER" OF A TRIANGLE? There's more than one way to define what we might mean by the center of a triangle, and each is charming in its own right.

Let's start with a point called them *centroid* of a triangle. In a triangle, draw line segments from each corner to the midpoint of the opposite side. Surprisingly, these three line segments intersect each other at a common point called the *centroid*; this is illustrated in this figure:

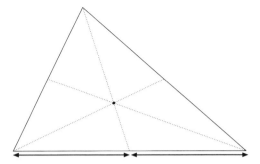

The centroid has the additional interesting property of being the center of mass of the triangle: if a triangle were made out of some sturdy material (say, a thin sheet of metal) the center of mass is the point at which the triangle would balance. Of course, the balance is fragile so the measurements would have to be precise.

INSTEAD OF DRAWING SEGMENTS TO THE OPPOSITE MIDPOINTS, let's draw the shortest possible segments from the corners of the triangle to the opposite side. In so doing, the line segments strike the opposing edge of the triangle at a 90° angle. Delightfully, these three segments also meet at a common point called the *orthocenter*. Here's a picture:

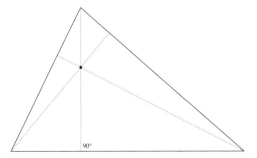

NEXT UP: ANGLE BISECTORS. Again, we draw three line segments emerging from the corners of the triangle, but this time the line segments are chosen to exactly bisect the angles. That is, the line segments divide the corner angles into two equal parts. As in the previous examples, these three segments also meet at a common point called the *incenter*.

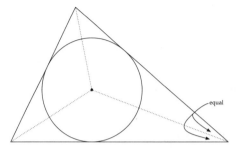

The incenter is so named because it is the center of the largest circle we can place inside the triangle; this is known as an *inscribed circle*.

INSTEAD OF DRAWING THREE SEGMENTS FROM THE CORNERS of the triangle, our next example features three segments drawn at right angles from the midpoints of the sides. Such segments are called *perpendicular bisectors* because they cut the sides into equal parts and are perpendicular to the sides. It's delightful to report that these three segments also meet at a common point called the *circumcenter* of the triangle. The name was chosen because this point is also the center of the smallest circle enclosing— *circumscribing*—the triangle.

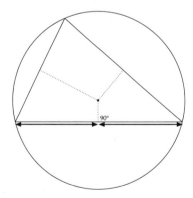

THESE FOUR DIFFERENT CENTERS (centroid, orthocenter, incenter, circumcenter) all land at the exact same location if the triangle is equilateral (all sides the same length). But, in general, they are all different. This picture shows (without the clutter of the various line segments) the four locations of these triangle centers.

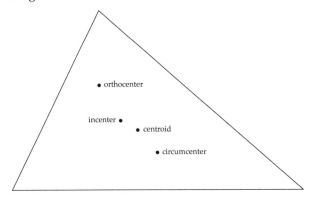

Interestingly, it's possible for some of these four "centers" to wind up *outside* the triangle! Can you figure out which? Answer on page 154.

Lurking equilateral triangles

INSTEAD OF ANGLE BISECTORS let's draw angle *trisectors* at each corner. That is, we draw two line segments emerging from each corner that divide the angle into three equal parts. This gives a total of six line segments (two from each corner). Of course, they don't all meet at a common point, but considering them in pairs (as in the picture) they mark the corners of a smaller triangle embedded in the larger.

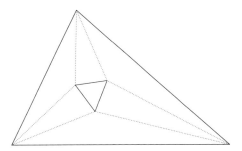

The stunning Morley's Theorem tells us that this

small triangle is always an equilateral triangle!

HERE'S ANOTHER WAY TO FIND an equilateral triangle lurking around any given triangle. Start with any triangle you please (shown thick in the diagram) and attach an equilateral triangle (drawn thin) to each of the three sides of the original triangle. We place a point in the center of each of these extensions, like this:

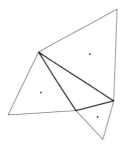

Next, connect the three centers and—voilà!—the new triangle is always an equilateral triangle (shown dotted in this picture):

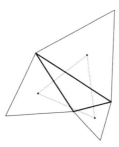

Pick's Theorem for quadrilaterals. Draw a quadrilateral on a grid and include one of its diagonals so we have two triangles glued together as shown in the figure on the facing page.

We can figure out the area of each triangle using Pick's Theorem and then add the two areas. For these two triangles (call them L and R for left and right) we have

$$I_L = 13 \quad B_L = 8 \quad \text{area} = 13 + \frac{1}{2} \times 8 - 1 = 16$$

$$I_R = 31 \quad B_R = 12 \quad \text{area} = 31 + \frac{1}{2} \times 12 - 1 = 36$$

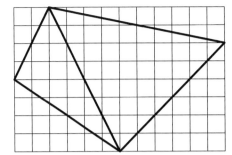

and so the area of the quadrilateral is $16 + 36 = 52$.

But delightfully, Pick's Theorem works unmodified for quadrilaterals! Here's why.

Instead of counting points anew, let's take advantage of the work we've already done.

There are 13 points interior to the left triangle and 31 in the interior of the right. But notice that there are 3 points on the diagonal that lie in the interior of the quadrilateral; we need to include them too. This gives $I_Q = 31 + 13 + 3 = 47$.

For the boundary of the quadrilateral, we know there are 8 grid points on the left triangle's boundary and 12 grid points on right's boundary, for a total of 20. But this is an overcount. The three interior grid points on the diagonal should never have been included and they were counted twice each; so we'll need to subtract 6 for them. And the two end points of the diagonal both got counted twice, so we need to subtract 2 to compensate. So $B_Q = 20 - 6 - 2 = 12$.

We now calculate

$$I_Q + \frac{1}{2} B_Q - 1 = 47 + \frac{1}{2} \times 12 - 1 = 52,$$

which is—amazingly—the right answer! What's going on?

The areas of the two triangles, L and R, add like this

$$\left(I_L + \frac{1}{2} B_L - 1 \right) + \left(I_R + \frac{1}{2} B_R - 1 \right)$$

to give the area of the quadrilateral. Let's rewrite this formula like this:

$$\text{area} = (I_L + I_R) + \frac{1}{2}(B_L + B_R) - 2.$$

The term $I_L + I_R$ misses some of the interior points in the quadrilateral. But the term $B_L + B_R$ overcounts the boundaries. The interior points on the diagonal are counted twice in this term, but really they belong to the I_Q term (and dividing by two exactly compensates). The ends of the diagonal are double counted in the boundary. Dividing by 2 half fixes the problem, but subtracting two (instead of one) sets everything perfectly right!

Indeed, Pick's Theorem works for any polygon whose corners lie on grid points.

Triangle centers outside the triangle. If the triangle is obtuse (one of its angles is greater than 90°) the circumcenter and the orthocenter will lie outside the triangle. Here is a picture in which the circumcenter lies outside the triangle.

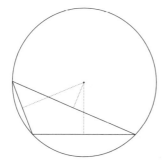

It's a bit trickier to locate the orthocenter of an obtuse triangle. The idea is to extend the sides of the triangle. Consider triangle ABC as drawn in this figure.

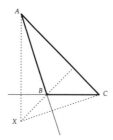

We locate the orthocenter by drawing these three lines: (1) a line through A perpendicular to BC (which we had to extend), (2) a line through B perpendicular to AC, and (3) a line through C perpendicular to AB (also extended). These three lines meet at X, the orthocenter.

14
Pythagoras and Fermat

The Pythagorean Theorem

THE SCARECROW DOESN'T GET A BRAIN at the end of *The Wizard of Oz*, but he does get a diploma. Proudly, he shows off his newfound intellect by utterly mangling the Pythagorean Theorem. He says:

> *The sum of the square roots of any two sides of an isosceles triangle is equal to the square root of the remaining side.*

In fact, the Pythagorean Theorem says nothing about isosceles triangles. Rather, it presents a relationship between the lengths of the sides of a *right triangle*: a triangle in which one of the angles is a right angle (90°).

An *isosceles* triangle is one in which two of the sides have the same length.

Given a right triangle, we use the letters a and b to stand for the lengths of the legs (the sides that form the right angle) and we use c for the length of the hypotenuse (the remaining side).

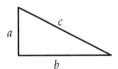

The Pythagorean Theorem asserts that these three lengths are related by the equation

$$a^2 + b^2 = c^2.$$

Here is how it is properly stated (as the Scarecrow undoubtedly intended):

Theorem (Pythagoras). *In a right triangle, the square of the length of the hypotenuse is equal to the sum of the squares of the lengths of the other two sides.*

The proof we present is based on the idea of geometric dissection: we make a figure, work out its area in two different ways, and—*voilà*—Pythagoras' Theorem emerges. Here's the proof.

START WITH FOUR IDENTICAL COPIES of a right triangle whose legs have lengths a and b and whose hypotenuse has length c. Take these four copies and arrange them to form a large $(a+b) \times (a+b)$ square as shown in this diagram:

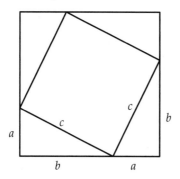

The area of this large square is $(a+b)^2 = a^2 + 2ab + b^2$.

Now we dissect the diagram into its five constituent pieces: four copies of the right triangle plus a $c \times c$ square. Here are the pieces rearranged for easy viewing:

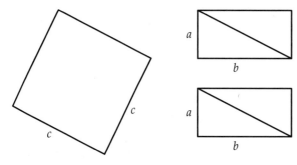

The Pythagorean Theorem shows that the diagonal of a 1×1 square has length $\sqrt{2}$. Here's how: Notice that the diagonal of such a square creates two right triangles whose legs are both length 1 and whose diagonal has unknown length, c. By the Pythagorean Theorem, $c^2 = 1^2 + 1^2 = 2$. Taking square roots gives $c = \sqrt{2}$. We discuss the number $\sqrt{2}$ in Chapter 4.

We've assembled the four right triangles into two rectangles to make the calculation of the total area as easy as possible: the square has area c^2 and the rectangles both have area ab. So the total area of these pieces is $c^2 + 2ab$.

We have determined the area of the figure in two ways: on the one hand the area is $a^2 + 2ab + b^2$ (from the original) and on the other it is $c^2 + 2ab$ (from the rearranged pieces). Since these are both correct ways to work out the area of the diagram, we conclude

$$a^2 + 2ab + b^2 = c^2 + 2ab.$$

Subtracting $2ab$ from both sides of this equation gives

$$a^2 + b^2 = c^2$$

and the Pythagorean Theorem is proved.

THERE ARE OTHER DISSECTION PROOFS of the Pythagorean Theorem that have the same flavor. We assemble some right triangles into a shape, work out the area of the shape, compare that to the areas of the constituent pieces, get an equation, and show that the algebraic relation $a^2 + b^2 = c^2$ follows.

For example, arrange the four right triangles into a $c \times c$ square like this:

This proof is attributed to Bhaskara, a twelfth-century Hindu mathematician.

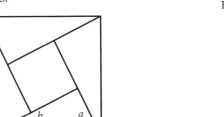

The total area of the figure is c^2. Now work out the area of the small, inner square plus the four triangles. Details appear on page 163.

Here is another dissection proof due to James Garfield, twentieth president of the United States.

Use two right triangles and a line segment to make a trapezoid.

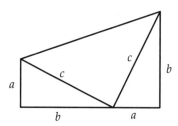

Work out the area of the trapezoid directly and then calculate it again by adding up the area of the three constituent triangles. Solutions appear on page 163.

A trapezoid is a four-sided figure in which two sides are parallel but the other two sides are not. The parallel sides are called the *bases* of the trapezoid. The formula for the area of a trapezoid is $\frac{1}{2}(b_1 + b_2)h$ where b_1, b_2 are the lengths of the bases and h is the distance between the bases.

Notice that Garfield's diagram can be seen as half of the diagram we used in our first proof; it's the result of cutting the middle $c \times c$ square on a diagonal.

Absolute value of complex numbers

The *absolute value* of a number is the result of dropping the number's minus sign (if it has one). For example, $|-5|$ is 5. We can think of a number such as -5 as being "five big" but in the negative direction. More rigorously, absolute value is defined like this:

$$|x| = \begin{cases} x & \text{if } x \geq 0 \text{ and} \\ -x & \text{if } x < 0. \end{cases}$$

So $|12| = 12$, $|-7| = 7$, and $|0| = 0$.

Here's a geometric interpretation: The absolute value of a number x is the distance between x and 0 on a number line.

This section deals with complex numbers, which we introduced in Chapter 5.

The absolute value of a number is how far the number is either to the left or to the right of 0; the sign of the number (positive or negative) doesn't matter.

How do we extend the idea of absolute value to complex numbers? What does $|3 + 4i|$ mean? We can't speak of $3 + 4i$ being positive or negative—those terms don't apply in this context. Instead, our goal is to define the absolute value of a complex number as the distance to 0. For this to work, we need a geometric picture of complex numbers. Just as real numbers are visualized as points on a line, complex numbers are represented as points in the plane. The complex number $3 + 4i$ is imagined to lie at a location 3 units to the right of the origin and then 4 units upward, as shown in the figure.

Visualizing the complex number $3 + 4i$ as a point in the plane.

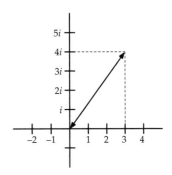

Now we have to work out the distance from $3 + 4i$ to the origin. This is indicated by the double headed arrow in the diagram. Notice that this arrow is the hypotenuse of a triangle whose legs have lengths 3 and 4. Let c be the length of this hypotenuse; the Pythagorean Theorem tells us that

$$c^2 = 3^2 + 4^2 = 9 + 16 = 25$$

and therefore $c = \sqrt{25} = 5$. Conclusion: $|3 + 4i| = 5$.

In general, the complex number $a + bi$ sits at a point that is a units horizontally and b units vertically from 0. A line segment joining the origin to $a + bi$ is the hypotenuse of a right triangle whose legs have lengths a and b. Using c to represent the length of this hypotenuse, the Pythagorean Theorem gives us that $c^2 = a^2 + b^2$ and so

$$|a + bi| = \sqrt{a^2 + b^2}. \tag{A}$$

Notice that this formula works for real numbers as well as for complex. For example, to calculate $|-4|$ the hard way, we think of -4 as $-4 + 0i$. Substituting $a = -4$ and $b = 0$ into equation (A) we have

$$|-4| = |-4 + 0i| = \sqrt{(-4)^2 + 0^2} = \sqrt{16} = 4.$$

Puzzle! The numbers 1, -1, i, and $-i$ all have absolute value equal to one, but they're not the only ones. Describe, geometrically, all the complex numbers whose absolute value is equal to 1. Answer on page 164.

Pythagorean triples

A RIGHT TRIANGLE, WHOSE LEGS HAVE LENGTHS 3 AND 4, has a hypotenuse whose length is 5. All three lengths are integers. Here's another example: If the

The 3-4-5 right triangle was known to the ancient Egyptians.

lengths of the legs are 5 and 12, then the hypotenuse length is

$$\sqrt{5^2 + 12^2} = \sqrt{25 + 144} = \sqrt{169} = 13,$$

which is also an integer. We're not always so lucky. If the leg lengths are 2 and 3, then the hypotenuse length is $\sqrt{13}$, which is not an integer.

Three positive integers, a, b, c, that are the lengths of the sides of a right triangle are called a *Pythagorean triple*. The simplest examples are $3, 4, 5$ and $5, 12, 13$. Are there others? How do we find them? Amazingly, the key to producing Pythagorean triples lies in the realm of complex numbers!

Before we dive into the details, let's see how the complex number $z = 2 + i$ yields the $3, 4, 5$ triple:

- Step 1: Calculate z^2. We have

$$z^2 = (2 + i) \times (2 + i) = (4 - 1) + (2 + 2)i = 3 + 4i.$$

- Step 2: Calculate $|z^2|$. Since $z^2 = 3 + 4i$ we have

$$|z^2| = \sqrt{3^2 + 4^2} = \sqrt{25} = 5.$$

The calculation in step 2 shows that $3, 4, 5$ is a Pythagorean triple: The line segment from 0 to $3 + 4i$ in the complex plane forms the hypotenuse of a right triangle whose sides have length 3 and 4, and the length of that segment is 5.

Let's try this procedure again with the complex number $z = 3 + 2i$. We calculate z^2 and then its absolute value:

$$z^2 = (3 + 2i) \times (3 + 2i) = (9 - 4) + (6 + 6)i = 5 + 12i$$
$$|z^2| = |5 + 12i| = \sqrt{5^2 + 12^2} = \sqrt{169} = 13.$$

And we've found the $5, 12, 13$ triple!

One more example and then we'll explore why this works. Start with $z = 5 + 2i$. Square this and calculate the absolute value of the result:

$$z^2 = (5+2i) \times (5+2i) = (25-4) + (10+10)i = 21+20i$$
$$|z^2| = |21+20i| = \sqrt{21^2 + 20^2} = \sqrt{441+400} = \sqrt{841} = 29.$$

We have found another Pythagorean triple: $20, 21, 29$.

LET'S EXPLORE WHY THIS WORKS by returning to the first example: $z = 2+i$. Notice that $|z| = \sqrt{2^2 + 1^2} = \sqrt{5}$. Now if we square z we get $z^2 = 3+4i$ and its absolute value is $|z^2| = \sqrt{3^2 + 4^2} = 5$. Summarizing:

$$|z| = \sqrt{5} \quad \text{and} \quad |z^2| = 5.$$

At least for this example, we have $|z^2| = |z|^2$.

Does $|z^2| = |z|^2$ always work? Certainly this is true for real numbers (for example, $|(-4)^2| = |16| = |-4|^2$) but verifying this for complex numbers requires a bit of algebra (worked out for you on page 163).

If you like, on first reading you can just trust that $|z^2| = |z|^2$ for all complex numbers. Later you can work out this bit of messy algebra.

Let's return to the procedure for creating Pythagorean triples. We begin with a complex number $z = x + yi$ where x and y are both integers. Its absolute value, $|z|$, might not be an integer, but it is the the square root of an integer: $\sqrt{x^2 + y^2}$. However, the absolute value of z^2 is necessarily an integer: $|z^2| = |z|^2 = x^2 + y^2$. Expanding z^2 we have

A complex number $x + yi$ where x and y are both integers is called a Gaussian integer.

$$z^2 = (x+yi) \times (x+yi) = (x^2 - y^2) + (2xy)i.$$

Taking $a = x^2 - y^2$, $b = 2xy$, and $c = x^2 + y^2$ we have that $|a+bi| = c$ and therefore $a^2 + b^2 = c^2$.

One last example: Let $z = 7 + 4i$. Its square is equal to $33 + 56i$, which has absolute value $|z^2| = \sqrt{33^2 + 56^2} = \sqrt{1089 + 3136} = \sqrt{4225} = 65$, and we have another Pythagorean triple: $33, 56, 65$.

We have shown that this procedure creates Pythagorean triples. It is natural to ask: Can every Pythagorean triple be created this way? The answer is yes, but the proof is more involved and we recommend consulting a book on number theory for more information.

Fermat's Last Theorem

WE HAVE JUST CONSIDERED TRIPLES OF INTEGERS THAT SATISFY the conclusion of the Pythagorean Theorem. That discussion was only marginally connected to the world of right triangles. We now leave geometry behind entirely and wonder about extensions of the relation $a^2 + b^2 = c^2$.

It's easy to find triples of integers a, b, c that satisfy the relation $a + b = c$. And the previous section gives a method for finding triples of integers that satisfy $a^2 + b^2 = c^2$. Our quest now is to extend to higher exponents: Can we find triples of integers that satisfy $a^3 + b^3 = c^3$? ... or $a^4 + b^4 = c^4$? ... or $a^5 + b^5 = c^5$, and so on?

Here are two uninteresting integer solutions to the equation $a^3 + b^3 = c^3$:

$$5^3 + 0^3 = 5^3 \quad \text{and} \quad 5^3 + (-5)^3 = 0^3.$$

The interesting challenge is to find three integers a, b, c *none of which is zero* for which $a^3 + b^3 = c^3$. Such a solution is called *nontrivial*.

This question was proposed by Pierre de Fermat in 1637. Writing in the margin of a book, Fermat asserts that the equation $a^n + b^n = c^n$ has no nontrivial integer solutions. He famously wrote (translated from Latin):

> *I have discovered a truly marvelous demonstration of this proposition that this margin is too narrow to contain.*

Because of his claim, this result is known as Fermat's Last Theorem, though it is doubtful that Fermat had a proof as it took more than three centuries for the

> This anecdote is sacred myth in the mathematics community and is recounted in many books and articles.
>
> Did Fermat have a proof? It is doubtful that he did. A more interesting question: Did Fermat *believe* he had a proof or was he joking? I prefer the latter interpretation.

matter to be settled by Andrew Wiles in the mid 1990s. Fermat's Last Theorem is, indeed, a theorem as Wiles' work shows that the equation $a^n + b^n = c^n$ has no nontrivial solution for any exponent $n \geq 3$.

In Bhaskara's diagram, the small square has area $(b-a)^2$. This gives the equation:
$$c^2 = (b-a)^2 + 2ab.$$
Note that $(b-a)^2 = a^2 + b^2 - 2ab$, so this equation becomes
$$c^2 = \left(a^2 + b^2 - 2ab\right) + 2ab = a^2 + b^2.$$

For Garfield's proof, the area of the trapezoid is
$$\frac{1}{2}(a+b)(a+b) = \frac{1}{2}a^2 + ab + \frac{1}{2}b^2$$
and the total area of the pieces is $ab + \frac{1}{2}c^2$ (ab for the two original right triangles and $\frac{1}{2}c^2$ for the other right triangle). Setting these equals gives
$$\frac{1}{2}a^2 + ab + \frac{1}{2}b^2 = \frac{1}{2}c^2 + ab.$$
Canceling ab from both sides and then multiplying by 2 yields $a^2 + b^2 = c^2$.

Proof that $|z^2| = |z|^2$.

The strategy is to consider the complex number $z = x + yi$ and then work out both $|z|^2$ and $|z^2|$; with luck, these will be the same.

We start with $|z|^2$. Remember that $|x + yi| = \sqrt{x^2 + y^2}$ and therefore
$$|z|^2 = \left(\sqrt{x^2 + y^2}\right)^2 = x^2 + y^2.$$

Next we work out z^2:
$$z^2 = (x+yi) \times (x+yi) = (x^2 - y^2) + (2xy)i.$$

Finally, we calculate $|z^2|$:
$$\begin{aligned}
\left|z^2\right| = \left|(x^2 - y^2) + (2xy)i\right| &= \sqrt{(x^2 - y^2)^2 + (2xy)^2} \\
&= \sqrt{(x^4 - 2x^2y^2 + y^4) + 4x^2y^2} \\
&= \sqrt{x^4 + 2x^2y^2 + y^4} \\
&= \sqrt{(x^2 + y^2)^2} \\
&= x^2 + y^2.
\end{aligned}$$

Notice that both $|z|^2$ and $|z^2|$ are equal to $x^2 + y^2$ and therefore they are equal to each other. This completes the proof that $|z|^2 = |z^2|$ for complex numbers z.

Solution to the puzzle (page 159): A complex number's absolute value equals its distance to the origin. So these numbers form a circle of radius one centered at the origin, $0 + 0i$.

15
Circles

CIRCLES ARE ELEGANT AND BEAUTIFUL. This chapter is a collection of some lovely facts about these fundamental geometric objects.

Mathematicians distinguish between a *circle*, which is a thin curve, and a *disk*, which is a filled-in region with a circle as its boundary.

A precise definition

MATHEMATICIANS DON'T USE SQUISHY DEFINITIONS; we require precision! A *circle* is a set of points in a plane a given distance from a given point. Let's tease that apart.

First, a circle is just a collection of points. Of course, not all collections of points form a circle. Only special ones. Which? A circle is a set of points determined by two inputs: a positive number r and a point X. The circle that these two inputs specify are those points at distance r from X. The point X is known, of course, as the *center* and the number r is the *radius*.

In a drawing of a circle (ink on paper or pixels on a screen) a circle has some thickness (or else it would be invisible) but mathematically circles are utterly thin.

Circles are close relatives of *spheres*: the set of points in three-dimensional space a given distance from a given point. Notice this definition is nearly the same; the only difference is that circles are con-

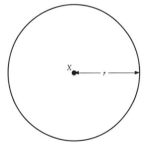

strained to lie in a plane.

An equation

POINTS IN THE PLANE can be specified by x, y-coordinates. When we are given an equation in the variables x and y, the set of points that satisfy that equation often determines a curve.

For example, the equation $x^2 + y^2 = 1$ is satisfied by some but certainly not all the points in the plane. For example, $(1, 0)$ satisfies this equation because $1^2 + 0^2 = 1$. Likewise, the point $(\frac{3}{5}, \frac{4}{5})$ also satisfied this equation; here's why:

$$\left(\frac{3}{5}\right)^2 + \left(\frac{4}{5}\right)^2 = \frac{9}{25} + \frac{16}{25} = \frac{25}{25} = 1.$$

On the other hand, the point $(\frac{1}{2}, \frac{1}{2})$ does *not* satisfy the equation because

$$\left(\frac{1}{2}\right)^2 + \left(\frac{1}{2}\right)^2 = \frac{1}{4} + \frac{1}{4} = \frac{1}{2} \neq 1.$$

Which points satisfy the equation $x^2 + y^2 = 1$? This is exactly the circle centered at $(0, 0)$ with radius 1.

Why? Think about a point (x, y). Use this point to make a triangle by drawing a vertical line segment from (x, y) to the horizontal axis, then from there another segment to the origin $(0, 0)$, and finally a third segment back to (x, y), as shown in the figure. The lengths of the legs of this right triangle are x and y, so by the Pythagorean Theorem (see Chapter 14) the length of the hypotenuse is $\sqrt{x^2 + y^2}$. That's the distance from the point (x, y) to the origin $(0, 0)$.

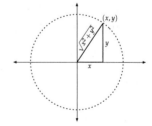

We only want those points that are distance 1 from the origin; that is, we require

$$\sqrt{x^2 + y^2} = 1.$$

Squaring both sides gives $x^2 + y^2 = 1$!

More generally, the circle with center at coordinates (a, b) and with radius r is given by the equation

$$(x - a)^2 + (y - b)^2 = r^2.$$

Triangles right inside

ANY TWO DISTINCT POINTS determine a straight line but three points might not be collinear. In that case, we can't draw a line through them, but there's exactly one circle that hits all three. We explained this in Chapter 13: for any triangle, the intersection of the three perpendicular bisectors of the sides meet at a point called the *circumcenter* of the triangle. That point is equidistant from the three corners of the triangle and so we can center a circle there that goes through the three points.

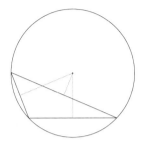

A question we might ask: When can a triangle be inscribed in a *semicircle*? In other words, we want one of the three sides of the triangle to be a diameter of the circle.

And there's a lovely answer: A triangle can be inscribed in a semicircle if and only if one of the angles is 90° (in other words, a right triangle).

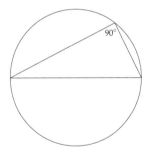

Ptolemy's Theorem

CHOOSE FOUR POINTS, $ABCD$, arranged cyclically on a circle. These four points determine six distances: The lengths of the sides of the quadrilateral, $|AB|$, $|BC|$, $|CD|$, and $|AD|$, as well as the lengths of the two diagonals d_1 and d_2; see the figure.

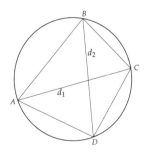

Ptolemy's Theorem gives an elegant, algebraic relation between these six quantities. Here it is:

$$d_1 \times d_2 = |AB| \times |CD| + |BC| \times |AD|.$$

What's more, suppose we have a quadrilateral whose sides and diagonals satisfy this algebraic relationship. Then the four corners of the quadrilateral lie on a common circle.

Packing

HOW TIGHTLY CAN CIRCLES BE PACKED? We're assuming here that all the circles have the same radius (say 1) and we want to pack as many circles as possible into a large planar region. (Imagine a huge tray that we wish to pack with a single layer of cans.)

One idea is to place them in a checkerboard-like pattern so each circle abuts four others, like this:

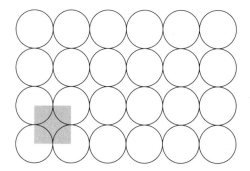

How efficient is this packing? One way to judge is to measure what fraction of the plane is covered by this pattern.

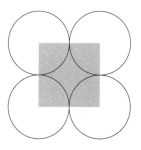

Look closely at four circles centered at the corners of a square. Since the circles have radius 1 the highlighted square has dimensions 2 × 2, so its area is 4. The square is not entirely covered by the four circles. Each of these four circles contributes one-fourth of its area, so all together the amount of the square covered by the four circles is the area of one full circle. Since the radius equals 1, that area is π.

That is, the fraction of the square covered is $\pi/4 \approx 0.785$. Since this pattern can be repeated in all directions to entirely cover the plane, we see that this packing covers 78.5% of the plane.

Not bad, but we can do better. Instead of aligning

the centers of the circles to a square grid, let's align them to the centers of a hexagonal grid like this:

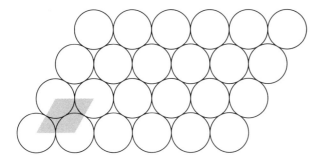

This packing covers more than 90% of the plane. We leave it as a geometric puzzle for you to figure out the exact amount; the answer appears on page 174.

THE HEXAGONAL PACKING OF CIRCLES in the plane is known to be the densest possible.

It's natural to ask: what happens in three dimensions? The answer—which may have been known since antiquity—was formally stated in the early 1600s by Johannes Kepler. Kepler asserted that the densest possible packing of spheres is achieved by stacking layers of hexagonal configurations (as shown in the figure on this page) atop each other so they fit as tightly as possible. This arrangement fills about 74% of the space.

The difficulty is to prove that no other packing of spheres can give higher density. While the situation in two dimensions is well established, a proof of Kepler's Conjecture proved elusive for nearly 400 years. Then, in the 1990s, Thomas Hales announced an extraordinary proof that combined both analytic reasoning with massive computation. Hales' proof has been scrutinized by experts and is generally believed to be correct.

Instead of packing cans in a flat tray, we are now filling a large box with identical balls. To see an illustration of the densest possible packing, look for a pyramid of oranges at your local grocery store.

The precise density of Kepler's sphere packing is $\pi/(3\sqrt{2})$.

Kissing circles

IF YOU DRAW THREE CIRCLES TANGENT TO EACH OTHER there's a small space in the middle where a fourth circle can be placed that is tangent to each of the big three. In this way, one can create an arrangement of four mutually tangent circles, like this:

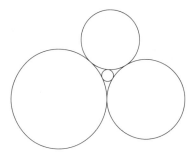

What relation is there between the sizes of the four circles? Or, asked differently, if we know the radii of three of the circles, can we use that information to figure out the radius of the fourth?

René Descartes considered exactly this problem and published a solution in the early 1600s. To present Descartes' solution in the simplest way, we define the *curvature* of a circle to simply be the reciprocal of its radius. In other words, a circle of radius 2 has curvature equal to $\frac{1}{2}$.

Here is Descartes' solution. If the curvatures of the four kissing circles are k_1, k_2, k_3, and k_4, then these numbers must satisfy this equation:

$$k_1^2 + k_2^2 + k_3^2 + k_4^2 = \frac{(k_1 + k_2 + k_3 + k_4)^2}{2}. \qquad (*)$$

For example, if the three large circles all have radius/curvature equal to 1 and the little circle that fits inside has curvature c, then equation $(*)$ requires that

$$3 + c^2 = \frac{(3+c)^2}{2}. \qquad (**)$$

Solving this quadratic equation gives $c = 3 \pm 2\sqrt{3}$.

Numerically, this is

$$c = 3 + 2\sqrt{3} \approx 6.464 \quad \text{and} \quad c = 3 - 2\sqrt{3} \approx -0.464.$$

The negative value doesn't apply (how can a circle have negative radius/curvature!?) and so we learn that the curvature of the small circle is about 6.464 and so its radius is $1/c \approx 0.1547$.

THERE'S ANOTHER WAY TO MAKE CIRCLES KISS. Again, start with three circles that are tangent to each other, but rather than inserting a small circle in the opening between them, make a big circle that hugs them from outside, like this:

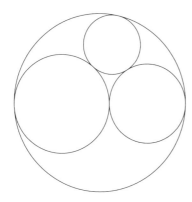

Delightfully, Descartes' solution still works. The trick is to think of the big circle as having a *negative* radius/curvature!

For example, if we begin with three pairwise tangent circles of radius/curvature equal to 1, then let c be the curvature of a larger circle that would just enclose them. Substituting $k_1 = k_2 = k_3 = 1$ and $k_4 = c$ into equation (∗) again yields equation (∗∗) with the same two solutions:

$$c = 3 + 2\sqrt{3} \approx 6.464 \quad \text{and} \quad c = 3 - 2\sqrt{3} \approx -0.464.$$

This time, however, the negative solution is the one that is relevant and tells us that the radius of the enclosing circle is approximately $1/0.464$, or 2.1547.

In other words, Descartes' equation works whether we are seeking the small circle that just fits inside the original three or a large circle that just encloses them. We simply provide the curvatures of the three original circles (as positive numbers) and solve for the fourth curvature. The positive result gives the curvature of the small circle and the negative result gives the curvature of the large one.

What we have done is use positive radius/curvature to signal that the tangency is exterior to the circle, while a negative radius/curvature indicates that the tangency is interior to the circle. This begs the question: What might *zero curvature* represent? As the words themselves suggest, a "circle" with zero curvature is simply a straight line.

DESCARTES' SOLUTION WAS REDISCOVERED in the 1930s by Frederick Soddy, who was so struck by the elegance of the solution that he wrote a poem entitled "The Kiss Precise" in celebration. Here is the second stanza of the poem in which we find Descartes' formula (∗) in rhyme:

> Four circles to the kissing come.
> The smaller are the benter.
> The bend is just the inverse of
> The distance from the center.
> Though their intrigue left Euclid dumb
> There's now no need for rule of thumb.
> Since zero bend's a dead straight line
> And concave bends have minus sign,
> The sum of the squares of all four bends
> Is half the square of their sum.

HERE IS ANOTHER WAY CIRCLES might kiss each other. As before, we have four circles, but this time they are tangent to each other in a ring. That is, the first and second are tangent, the second and third are tangent, the third and fourth are tangent, and then the fourth and first are tangent. Such an arrangement creates exactly four points of tangency where consecutive circles meet.

Here's the surprising conclusion: Those four points of tangency always lie together on a common circle, like this:

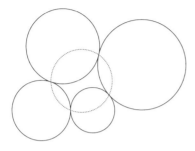

Pascal's Hexagon Theorem

WE CLOSE THIS CHAPTER with a result of Blaise Pascal.

Place six points A, B, C, D, E, F around a circle. We use them to make a twisted hexagon by connecting them with line segments as follows:

$$A \to D \to B \to F \to C \to E \to A.$$

As these segments crisscross from one side of the circle to the other, three intersections will be formed; let's call those points X, Y, and Z.

Pascal's Hexagon Theorem asserts that those three points of intersection, X, Y, Z, will always lie on a straight line! Here's an illustration:

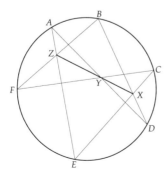

174   THE MATHEMATICS LOVER'S COMPANION

And, as an added bonus, Pascal's Hexagon Theorem also works if the six points are placed around an ellipse.

The density of the hexagonal circle packing.
  Suppose that the circles have radius 1. Notice that the four circles shown in the figure are centered at the corners of a rhombus whose sides have length 2.

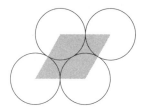

This rhombus can be thought of as two equilateral triangles (draw the short diagonal) glued together.
  An equilateral triangle with side length 2 has altitude $\sqrt{3}$. Therefore the area of such a triangle is

$$\frac{1}{2} \times \text{base} \times \text{height} = \frac{1}{2} \times 2 \times \sqrt{3} = \sqrt{3}$$

and so the area of the shaded rhombus is $2\sqrt{3}$.
  Now look at the shaded parts of the four circles. Two are one-sixth shaded and the other two are one-third shaded. All together, the shaded area amounts to one entire circle. Since we assumed the circles have radius 1, that total area is π.
  It now follows that the fraction of the rhombus covered by the circles is

$$\frac{\pi}{2\sqrt{3}},$$

which is about 90.7%.

An altitude of such an equilateral triangle cuts the triangle into two right triangles with a hypotenuse of length 2 and one leg of length 1. The Pythagorean Theorem gives us the length of the altitude.

16
The Platonic Solids

AN EQUILATERAL TRIANGLE IS A THREE-SIDED SHAPE composed of three equally long line segments that meet in 60° angles. A square is a four-sided figure composed of four equally long line segments that meet in 90° angles. These are examples of *regular polygons*: figures bounded by equal-length line segments that meet in equal-size angles. The diagram shows a regular heptagon (7-gon). The shape of a stop sign is a regular octagon (8-gon).

This is a regular 7-gon; its sides all have the same length and its angles are all the same size.

Puzzle: The corners of a regular 7-gon all have the same size. How large is that angle? Hint: Remember that the sum of the angles of a triangle is 180°. The answer is on page 189.

A moment's reflection reveals that there are infinitely many types of regular polygons: There's a regular n-gon for every integer $n \geq 3$.

Polygons are figures drawn in the plane. What is the analogous situation for figures drawn in three-dimensional space?

Polyhedra

THE GENERALIZATION OF A POLYGON IN THREE-DIMENSIONAL SPACE is called a *polyhedron*. This is a solid figure with flat sides each of which is a polygon. Among the more familiar polyhedra are the triangular prism and the square-base pyramid. The prism is made of three rectangles and two triangles. The pyramid consists of four triangles and a square.

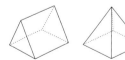

How do we generalize the idea of a regular polygon to three-dimensional figures? Regular polygons have sides, and angles, that are congruent to each other.

The extension to three dimensions is to require that all "parts" of a polyhedron be congruent to each other. That is, we require:

- all edges of the polyhedron are congruent to each other,
- all the angles formed where two edges meet are congruent,
- the number of edges meeting is the same at every vertex (corner), and
- all angles between faces that share an edge are congruent.

Line segments that are the same length or angles that are the same measure are called *congruent*. The idea of congruence applies beyond segments and angles. Any two figures that are precisely the same shape are deemed congruent.

The first two requirements imply that the faces of a regular polyhedron be congruent, regular polygons.

Perhaps the best-known regular polyhedron is the *cube*, which consists of six faces all of which are regular 4-gons (squares). In addition to the cube, the figure shows four other regular polyhedra.

cube   tetrahedron   octahedron   dodecahedron   icosahedron

- The *tetrahedron* is built from four equilateral triangles.
- the *octahedron* consists of eight equilateral triangles. (Imagine two square-base pyramids glued together.)
- the *dodecahedron* is built from twelve regular pentagons, and
- the *icosahedron* is built from twenty equilateral triangles.

The following diagrams show the regular polyhedra flattened into the plane. You can try copying these pictures, cutting them out, folding on the lines, and taping together to make paper models; there are commercially available kits for doing exactly this.

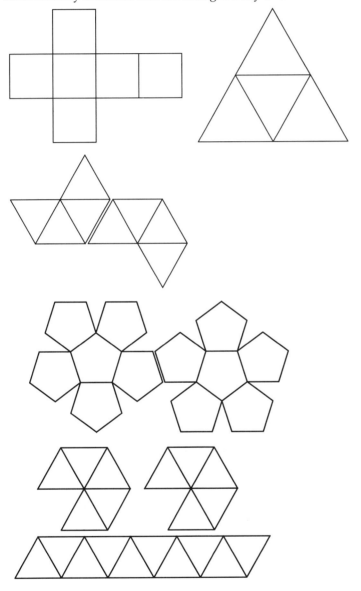

These five regular polyhedra are known as the *Platonic solids*. Are there other regular polyhedra?

The figure at the right shows a *stellated icosahedron* whose faces are all equilateral triangles, but it is not a regular polyhedron because the angles between the faces are not all the same and the number of edges meeting at the vertices varies (three at the extreme corners and ten at the nestled vertices).

Aiding us in our hunt for additional regular polyhedra is a lovely formula due to Leonhard Euler (whom we introduced in Chapter 7).

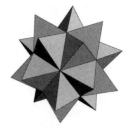

stellated icosahedron

Euler's polyhedral formula

A POLYGON HAS THE SAME NUMBER OF VERTICES (corners) as it does edges. The situation for a polyhedron is more complicated as they are comprised of vertices, edges, and faces. This chart gives a parts census for all of the polyhedra we've considered thus far:

Polyhedron	Vertices	Edges	Faces
Prism	6	9	5
Pyramid	5	8	5
Cube	8	12	6
Tetrahedron	4	6	4
Octahedron	6	12	8
Dodecahedron	20	30	12
Icosahedron	12	30	20
Stellated Icosahedron	32	90	60

As you pore over this chart looking for patterns, you may notice that the data for the cube and the octahedron are reverses of one another: $(8, 12, 6)$ versus $(6, 12, 8)$. The same reversal holds for the dodecahedron/icosahedron: $(20, 30, 12)$ vs. $(12, 30, 20)$. These reversals hold because of a phenomenon known as *duality*.

If you place a point at the center of each of the six faces of a cube and then connect dots from faces that share an edge, a smaller polyhedron emerges inside the cube: it's an octahedron. Conversely, if you place a point in the center of each triangular face of an octahedron and connect the dots, a cube is formed. The same duality holds for the icosahedron and dodecahedron.

Look closely at this chart to see if you can discover a simple relation among the number of vertices, edges, and faces for a polyhedron. The answer is revealed just below, but it's more fun if you can stumble on the equation yourself. Use the variable names V, E, and F for the number of vertices/edges/faces, respectively.

WHILE YOU'RE CONTEMPLATING the relation among V, E, and F, let's pause to check the entries in the table. For a simple solid, such as the pyramid, it's not difficult to count the parts. There are five vertices (four around the base and one at the apex), eight edges (four around the base and four leading up to the top), and five faces (four triangles, one square). The tetrahedron and prism are easy to check. The cube is not too challenging because it is such a familiar shape. There are 8 vertices (four on the top square, four on the bottom), 12 edges (four around the top, four around the bottom, and four verticals), and 6 faces (we've all played with dice).

The other polyhedra are more difficult to visualize. It helps to squash them flat like this: Imagine the polyhedron is hollow and we use a scissor to remove one of the faces. We then stretch that empty face wide until the ball becomes flattened. The results look like these pictures:

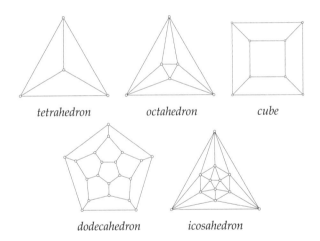

tetrahedron   octahedron   cube

dodecahedron   icosahedron

Let's start with the octahedron. From the picture, it's easy to see that $V = 6$. Counting faces, it's easy to make the mistake that there are seven, but remember there's a face we removed before we flattened, so $F = 8$.

Here's a trick to make counting the edges easier. For every vertex in the diagram, place little tick marks near each corner across every edge that touches that corner, like this:

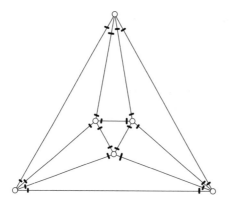

How many tick marks have we laid down? Since there are four edges meeting at each vertex, the number of tick marks is four times the number of vertices: $4 \times V = 4 \times 6 = 24$. On the other hand, every edge has exactly two tick marks, so the number of ticks is $2 \times E$, so E must be 12.

Let's try this same idea for the icosahedron. Looking at the flattened picture, we see three vertices at the extreme corners, six more that form a hexagon, and then three more in the center giving $V = 3 + 6 + 3 = 12$. Counting faces, there are 9 triangles that reach the extreme corners, another ring of 9 inside the pentagon, and then one more in the very center. That gives $9 + 9 + 1 = 19$, but there is also the face we removed, so we have $F = 20$. Finally, to count edges we use the same "tick trick" from before. If we place ticks on edges near each vertex, we'll place a total of $5 \times 12 = 60$ ticks: five ticks near each of the twelve vertices. Since every edge gets two ticks this way, we find that $E = 30$.

IT'S TIME FOR US TO REVEAL the lovely relation among the number of vertices V, edges E, and faces F of a polyhedron that was first derived by Euler and (we hope) rediscovered by you.

Notice that the number of vertices plus faces, $V + F$, is always two larger than the number of edges.

For example, the cube has $V = 8$ and $F = 6$ for a total of $V + F = 14$, which is two larger than $E = 12$. We can write this relation as $V + F = E + 2$, but it has become somewhat traditional to write Euler's formula like this:

$$V - E + F = 2. \qquad (A)$$

Let's see why this works.

We start by flattening the polyhedron into the plane by removing a face and stretching. The number of regions in the flattened picture is exactly equal to the number of faces F because the removed face corresponds to the region surrounding the entire picture; all the other faces correspond to finite regions. The numbers of vertices, edges, and regions in this picture are precisely V, E, and F, respectively. The algebraic expression $V - E + F$ has some numerical value; I've got to convince you that it's 2.

The flattened picture of a polyhedron is an example of a *graph*—the mathematical abstraction of a network.

To convince you that $V - E + F = 2$, I begin by erasing an edge. What happens to the number of vertices, edges, and regions as a result? The number of vertices doesn't change—all I've done is erase an edge. The number of edges goes down by 1, of course. What happens to the number of faces? As you can see in the picture, the two faces on either side of the doomed edge combine to form a single face, so the number of faces also goes down by one. If the new picture has V', E', F' vertices/edges/regions, then these updated counts are related to the originals like this:

$$V' = V, \qquad E' = E - 1, \qquad \text{and} \qquad F' = F - 1.$$

It follows that $V' - E' + F' = V - (E - 1) + (F - 1) = V - E + F$. If I can convince you that $V' - E' + F' = 2$, then I've demonstrated that $V - E + F = 2$.

Here's the strategy:

I'm going to keep deleting edges from the picture. Every time I do, I lose an edge, but I also lose a region (because two regions combine). But I need to be careful. As I'm deleting edges, I might run into a situation in which the two sides of an edge are part of the boundary of the same region; see the thick edge in the figure. Such an edge cannot be part of a cycle of edges in the picture because then the cycle would separate the plane into the regions inside the cycle and the regions outside the cycle. So I'm safe deleting edges as long as they are part of a cycle. Every time I do, I decrease the number of edges and the number of faces by one, so the value of $V - E + F$ (whatever it is) doesn't change as the result of such a deletion.

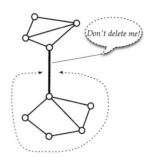

Eventually I will come to a point when there are no cycles left in the picture. All of the regions have been coalesced into one (so, at this point, $F = 1$) and there are no edges that I can "safely" remove. (See the figure.) We now switch to phase two of the seek-and-destroy mission.

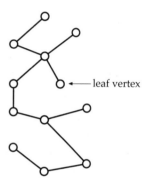

leaf vertex

The picture is now without cycles. Pick a vertex, any vertex, and start following a path along edges and vertices from that starting location. Because there are no cycles, this path cannot return to a previously visited vertex; but neither can this path go on forever (there are only finitely many vertices). So the path must eventually get "stuck" at a vertex that is the end point of just a single edge. Such a vertex is called a *leaf*. Our next step is to delete a leaf vertex and the edge on which it hangs. What happens to the value of $V - E + F$? The number of vertices goes down by 1 (the leaf we delete), the number of edges goes down by 1 (the single edge touching that vertex), and the number of faces is unchanged (still 1). Writing V', E', F' for the new counts after deleting the vertex and edge, we have

$$V' = V - 1, \quad E' = E - 1, \quad \text{and} \quad F' = F = 1.$$

Therefore $V' - E' + F' = (V - 1) - (E - 1) + F = V - E + F$. Whatever value $V - E + F$ had before the leaf/edge deletion, the new value is unchanged.

After deleting a leaf and its edge from the picture, we are left with a new picture that also has no cycles. So we find another leaf, delete it and its associated edge, and repeat. We keep doing this until there is only one vertex left in the graph. Each vertex/leaf deletion leaves the value of $V - E + F$ unchanged.

Recapping: We flatten the polyhedron. We delete edges that are a part of a cycle until there are no cycles left; although the numbers V, E, or F might change, the value of $V - E + F$ does not. After all the cycles are gone, we repeatedly find a leaf and its associated edge until the entire picture is reduced to a single vertex. Again, although the individual values V, E, and F might change, the value of $V - E + F$ does not. At the very end, we have a single vertex, no edges, and a single region (the infinite region surrounding the lone vertex). That is, at the end we have $V = 1$, $E = 0$, and $F = 1$; for these numbers, we have that $V - E + F = 2$. Since the deletions didn't affect the overall value of $V - E + F$, the original picture must obey the relation $V - E + F = 2$. This completes our justification of Euler's polyhedral formula (A)!

Is that all there is?

WE HAVE PRESENTED FIVE REGULAR POLYHEDRA: the Platonic solids tetrahedron, cube, octahedron, dodecahedron, and icosahedron. With the help of formula (A), we'll show that this is the complete list; there are no more.

We will use five numbers to describe the regular polyhedra. The first three are the now-familiar V, E, and F—the number of vertices, edges, and faces, respectively. Every face of a regular polyhedron is a regular polygon; we'll use n to stand for the number of sides on each face. The number of edges incident at a vertex of a regular polyhedron is the same at every vertex; we'll use r to stand for that number.

Here's the data for the five Platonic solids:

Polyhedron	V	E	F	n	r
Tetrahedron	4	6	4	3	3
Cube	8	12	6	4	3
Octahedron	6	12	8	3	4
Dodecahedron	20	30	12	5	3
Icosahedron	12	30	20	3	5

Here's a summary of the five numbers for easy reference:
- V: number of vertices
- E: number of edges
- F: number of faces
- n: number of edges surrounding a face
- r: number of edges meeting at a vertex

Now we work out some algebraic relations among these quantities. The first is Euler's formula, which we repeat here:

$$V - E + F = 2. \qquad (A)$$

Second, we use the tick trick to find a relation among E, V, and r. We place a small tick mark on every edge near its end points. In this way, every edge gets two ticks (one on each end) so the number of ticks placed is $2E$. On the other hand, we place r ticks near each of the V vertices; therefore we have laid down rV ticks. As these are both correct accounting for the number of ticks, they must equal each other:

$$2E = rV. \qquad (B)$$

Third, we use the tick trick again, but this time the ticks come from the *faces*. That is, we go around each face and place a tick on the edges that surround it. As before, every edge gets two ticks (one from each side). On the one hand, the number of ticks is $2E$ (because every edge has two ticks), but on the other hand, the number of ticks is nF (n ticks from each of F faces). Therefore:

$$2E = nF. \qquad (C)$$

Let's check equations (A), (B), and (C) for the

dodecahedron:

$$V - E + F = 20 - 30 + 12 = 2, \qquad \text{(A)}$$
$$2E = 2 \times 30 = 60 = 3 \times 20 = rV, \quad \text{and} \quad \text{(B)}$$
$$2E = 2 \times 30 = 60 = 5 \times 12 = nF. \qquad \text{(C)}$$

From (B) we have $V = 2E/r$ and from (C) we have $F = 2E/n$. Substituting these into (A) gives

$$V - E + F = 2$$
$$\frac{2E}{r} - E + \frac{2E}{n} = 2 \qquad \text{substitution}$$
$$\frac{1}{r} - \frac{1}{2} + \frac{1}{n} = \frac{2}{E} \qquad \text{divide by } 2E$$
$$\frac{1}{r} + \frac{1}{n} = \frac{1}{2} + \frac{1}{E} \qquad \text{add } \frac{1}{2} \qquad \longleftarrow \text{we'll use this equation later}$$

and since $\frac{1}{2} + \frac{1}{E} > \frac{1}{2}$ we conclude

$$\frac{1}{r} + \frac{1}{n} > \frac{1}{2}. \qquad \text{(D)}$$

Relation (D) implies that r and n cannot be too large. For example, we can't have $r = n = 5$, because then we'd have $\frac{1}{r} + \frac{1}{n} = \frac{1}{5} + \frac{1}{5} = \frac{2}{5}$, which isn't greater than $\frac{1}{2}$. Let's work out the possible values for r and n.

First note that n and r must be at least 3. The faces are n-gons and the smallest polygon is a triangle; hence $n \geq 3$. The polyhedron is a solid and if $r = 2$ there would only be two edges coming together at a vertex; we need at least $r \geq 3$ in order for there to be any thickness.

Now let's step through the possible values of n.

- $n = 3$. By (D), $\frac{1}{3} + \frac{1}{r} > \frac{1}{2}$. Subtracting $\frac{1}{3}$ from both sides gives $\frac{1}{r} > \frac{1}{2} - \frac{1}{3} = \frac{1}{6}$, which implies that $r < 6$.

So if $n = 3$, the only possible values for r are 3, 4, and 5.

- $n = 4$. By (D), $\frac{1}{4} + \frac{1}{r} > \frac{1}{2}$, which yields $\frac{1}{r} > \frac{1}{4}$ or $r < 4$. The only possible value for r in this case is 3.

- $n = 5$. By (D), $\frac{1}{5} + \frac{1}{r} > \frac{1}{2}$, which gives $\frac{1}{r} > \frac{1}{2} - \frac{1}{5} = \frac{3}{10}$ and hence $r < \frac{10}{3} = 3\frac{1}{3}$. This implies that $r = 3$.

- Finally, $n \geq 6$, so $\frac{1}{n} \leq \frac{1}{6}$. By (D), $\frac{1}{n} + \frac{1}{r} > \frac{1}{2}$, which gives $\frac{1}{r} > \frac{1}{2} - \frac{1}{6} = \frac{1}{3}$. This implies that $r < 3$, which is impossible. Hence n cannot be 6 or larger.

Summarizing, there are only five possibilities for the pair of values (n, r):

$$(3,3), \quad (3,4), \quad (3,5), \quad (4,3), \quad \text{and} \quad (5,3).$$

Given the values of n and r, we can work out the value of E (using the equation $\frac{1}{r} + \frac{1}{n} = \frac{1}{2} + \frac{1}{E}$) and then we can derive V and F using equations (B) and (C). Here are the calculations in the five cases:

- $(n, r) = (3, 3)$: From $\frac{1}{r} + \frac{1}{n} = \frac{1}{2} + \frac{1}{E}$ we have $\frac{1}{3} + \frac{1}{3} = \frac{1}{2} + \frac{1}{E}$, which gives $\frac{1}{E} = \frac{2}{3} - \frac{1}{2} = \frac{1}{6}$ and so $E = 6$.

    By (B), $2E = rV$ we get $12 = 3V$ and so $V = 4$.

    By (C), $2E = nF$ we get $12 = 3F$ and so $V = 4$.

    Conclusion: $(n, r) = (3, 3)$ yields $(V, E, F) = (4, 6, 4)$. The only way to assemble $F = 4$ equilateral triangles ($n = 3$) into a solid is the tetrahedron.

- $(n, r) = (3, 4)$: From $\frac{1}{r} + \frac{1}{n} = \frac{1}{2} + \frac{1}{E}$ we have $\frac{1}{3} + \frac{1}{4} = \frac{1}{2} + \frac{2}{E}$ hence $\frac{1}{E} = \frac{1}{3} + \frac{1}{4} - \frac{1}{2} = \frac{1}{12}$ giving $E = 12$.

    By (B), $2E = rV$ we get $24 = 4V$ and so $V = 6$.

    By (C), $2E = nF$ we get $24 = 3F$ and so $F = 8$.

    Conclusion: $(n, r) = (3, 4)$ yields $(V, E, F) = (6, 12, 8)$. The only way to assemble $F = 8$ equilateral triangles ($n = 3$) into a solid (with $r = 4$ triangles at each corner) is the octahedron.

- $(n, r) = (4, 3)$. The calculations are nearly identical to the $(3, 4)$ case and they yield $(V, E, F) = (8, 12, 6)$. The only way to assemble $F = 6$ squares ($n = 4$) into a solid with $r = 3$ meeting at each corner is the cube.

- $(n, r) = (3, 5)$. From $\frac{1}{r} + \frac{1}{n} = \frac{1}{2} + \frac{1}{E}$ we have $\frac{1}{3} + \frac{1}{5} = \frac{1}{2} + \frac{1}{E}$ hence $\frac{1}{E} = \frac{1}{3} + \frac{1}{5} - \frac{1}{2} = \frac{1}{30}$ giving $E = 30$.

  By (B), $2E = rV$ we get $60 = 5V$ and so $V = 12$.

  By (C), $2E = nF$ we get $60 = 3F$ and so $F = 20$.

  Conclusion: $(n, r) = (3, 5)$ yields $(V, E, F) = (12, 30, 20)$. The only way to assemble $F = 20$ equilateral triangles ($n = 3$) with $r = 5$ meeting at every corner is the icosahedron.

- $(n, r) = (5, 3)$. The calculations are nearly identical to the $(3, 5)$ case and they yield $(V, E, F) = (20, 30, 12)$. The only way to assemble $F = 12$ regular pentagons with three meeting at each corner is the dodecahedron.

Thanks to Euler's remarkable formula, and a good bit of algebra, we have demonstrated that the five Platonic solids are the only regular polyhedra!

Archimedian solids

THE FACES OF A REGULAR POLYHEDRON MUST ALL BE IDENTICAL regular polygons, but if we relax that restriction somewhat a new category of polyhedra emerges. We still ask that the faces be regular polygons, but we allow a mix of different types of polygons. Instead, we impose a symmetry requirement that the polyhedron looks "the same" at every vertex. We call such polyhedra *semiregular*.

For example, we can make a prism from two equilateral triangles and four squares. Every corner of the prism looks exactly like every other: it is the junction of two squares and a triangle.

We can make prisms with other shaped ends. For example, we can connect two parallel regular pentagons by five squares. In this way, there's an entire infinite family of semiregular polyhedra.

Here's another infinite family: Take two regular n-gons (for example, pentagons) parallel to each other, but with one rotate half a "click" from the other. Connect the points on one with the points on the other in a zig-zag fashion forming a ring of triangles. If we adjust the distance between the two layers just right, those triangles will be equilateral. Polyhedra formed in this manner are known as *antiprisms*.

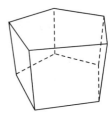

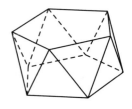

One of the Platonic solids is a prism and another is an antiprism. Can you figure out which? The answer is on the facing page.

THE PRISMS, ANTIPRISMS, AND PLATONIC SOLIDS are not the only semiregular polyhedra. There are, in addition, thirteen *Archimedian solids*. A complete accounting of these polyhedra can be found elsewhere; here we'll mention just one favorite.

If one slices off a corner of an icosahedron, the cross section will be a regular pentagon because there are five triangles meeting at each vertex. If we do this at all twelve corners, then the 20 triangular faces become hexagons. If the cuts are made at precisely the right depth, then the lengths of the sides of the hexagon will all be equal. The result is a *truncated icosahedron*. If we were to create a physical model of a truncated icosahedron from sturdy fabric, color the hexagons white, the pentagons black, and inflate with air, the result is the familiar soccer ball!

Solution to the puzzle on page 175: The line segments of a regular 7-gon meet to form $128\frac{4}{7}$-degree angles. Here's why: Dissect the regular 7-gon into seven isosceles triangles meeting at the center. The peaks of these isosceles triangles are of size $360 \div 7 = 51\frac{3}{7}$ degrees. Let x be the base angle of the isosceles triangle and we have $x + x + 51\frac{3}{7} = 180$. Solving gives $x = 64\frac{2}{7}$. The angles at the corners of the 7-gon have size $2x$, which is $128\frac{4}{7}$ degrees.

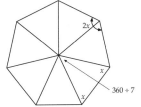

Here is an alternative solution: The sum of the angles of an n-gon is $180(n-2)$. In the case of a regular n-gon, all the angles are the same, so each has size $180(n-2)/n$. Substituting $n = 7$ gives $180 \times 5/7 = 900/7 = 128\frac{4}{7}$.

Platonic prism/antiprism: The cube is a prism formed by using two squares for the bases, and the octahedron is the antiprism created from two equilateral triangles.

17
Fractals

THE SHAPES WE MEET IN GEOMETRY CLASS are simple and pure. They have clean, sharp edges. Line segments are perfectly straight and circles are exquisitely round. Viewed from space, the Earth appears to be the smoothest marble we've encountered, but up close it's a different story. Craggy mountain peaks tower over ripply sand dunes and wavy oceans. Rivers are meandering, branching structures lined with leafy trees. If we ask an artist to render a natural scene with only line segments and circular arcs, the best we can hope for is an inspired abstraction.

The shapes that nature produces often have rough or ill-defined edges. What is the "shape" of a cloud or of a raging fire? Euclid's *Elements* is of little help. A different category of shapes is needed to describe our crunchy, fuzzy world.

Sierpinski's triangle

WE BEGIN WITH A RECIPE.

We need a piece of dough and several extremely sharp knives. We're also going to employ a hoard of chefs.

The piece of dough is a perfect equilateral triangle. The master chef carefully cuts out (and discards) the equilateral triangle formed by joining the midpoints of the sides. The process is illustrated in the figure.

What remains are three equilateral triangles—each half the size of the original—joined at their corners.

Now the master chef summons three sous chefs and orders them to do as exactly as she did: Cut out the middles of the three smaller triangles. The resulting piece of dough consists of nine quarter-size triangles.

Of course, the sous chefs hope someday to be master chefs, so each summons three sous-sous chefs to repeat the process on the quarter-sized triangles.

And the process continues. Each of the sous-sous chefs hands off his or her triangle to three sous-sous-sous chefs and orders them to carefully excise the middle portion of their triangles. The steps looks like this:

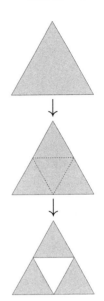

The delegating and cutting continues to infinitely many minions mincing merrily away. At the end—whatever that means!—we have created *Sierpinski's triangle*, which looks like this:

Precisely speaking, the removal process deletes the *interiors* of the triangles; the bounding line segments are not removed and, at the end, are all that remain.

SIERPINSKI'S TRIANGLE HAS TWO SALIENT PROPERTIES that earn it the name *fractal*: it is *self-similar* and has *fractional dimension*.

The notion of self-similarity is easier to understand. When you look at Sierpinski's triangle, what you see are three half-size copies of itself. And each of those copies is made up of three smaller (quarter-size) copies. Indeed, if you take a powerful microscope and zoom in on any tiny region, you'll find a miniature replica of the full figure. Each piece is exactly the same as the whole, only smaller.

Between dimensions

THE OBJECTS OF EUCLIDEAN GEOMETRY can be neatly sorted into categories based on their *dimension*.

Line segments, arcs of circles, boundaries of squares, and the like are one dimensional. They have length but take up no area. A helical coil is one dimensional even though it extends into three-dimensional space.

When we include their interiors, rectangles, pentagons, circles, and so on are two dimensional: they have area but take up no volume. The surface of a cylinder is two dimensional even though it doesn't lie flat in a plane.

And spheres, cubes, and so on (when we include their interiors) are three dimensional; they have volume.

But what about Sierpinski's triangle? It starts off as an equilateral triangle so perhaps it is two dimensional. In that case, what is its area?

For simplicity, suppose the area of the initial piece of dough measures 1 unit of area (e.g., one square centimeter). The dotted lines divide the initial triangle into four identical pieces. The discarded piece of dough has area $\frac{1}{4}$, leaving $\frac{3}{4}$ of the original for the next round.

Next the three sous chefs make their cuts to remove the middles of their quarter-sized pieces. Their efforts remove one-fourth of the remaining area; as before, that means that $\frac{3}{4}$ is preserved for the next step. The sous-sous chefs remove a quarter of what they received, diminishing the area again by a factor of $\frac{3}{4}$. In other words, after n rounds of cutting, the amount of dough left from the original is $\left(\frac{3}{4}\right)^n$.

After 16 steps of this process, 99% of the area has been removed. As n tends to infinity, we find that *all* of the area has been eliminated; all that remains are the bounding line segments and they cover no area whatsoever.

Is Sierpinski's triangle one dimensional?

Let's try to calculate its length.

An equivalent way to create Sierpinski's triangle is to begin with an *empty* equilateral triangle. We then add three line segments joining the midpoints of the three sides. We then repeat that operation on the three newly formed triangles, but not the one in the middle. And we repeat again and again like this:

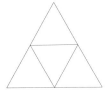   

For convenience, let's assume that the sides of the original triangle are all 1 unit long. The total length of the line segments in the original triangle is 3.

When the first inner triangle is inserted the three new line segments have length $\frac{1}{2}$, so this first insertion adds $\frac{3}{2}$ to the total length.

At the second step we add nine line segments (three in each of the three triangles). Each of these has length $\frac{1}{4}$, so the additional length is $\frac{9}{4}$.

Step three adds 27 line segments (three in each of nine triangles). These segments have length $\frac{1}{8}$, so this

Note: If the sides of an equilateral triangle have length equal to 1, its area is not 1. We're starting with a different sized triangle to make the calculations cleaner.

step adds $\frac{27}{8}$ to the total.

The next step adds 81/16, and so on. The additional length added at step n is $\left(\frac{3}{2}\right)^n$. Notice that as n grows, this amount gets larger and larger.

Conclusion: The total length of all the line segments in Sierpinski's triangle is infinite!

Sierpinski's triangle has zero area and infinite length. In this sense, it is an object that's "bigger" than one dimensional but "smaller" than two. But this is vague; we can be much more precise. We're going to show that its dimension is 1.5849625007.... Really.

Box counting

THE DIMENSION OF A GEOMETRIC FIGURE is a measure of its "thickness." A one-dimensional object, such as a line segment, is "thinner" than a filled-in triangle that, in turn, is "thinner" than a solid ball. Let's see how we can make this vague idea of "thin" or "thick" into a mathematically precise quantity.

The idea is to draw the figure on graph paper. Or, more precisely, we repeatedly draw it on graph paper, but in each subsequent drawing we use a finer and finer grid.

Let's illustrate this idea with a simple squiggle. That is, we draw the same curve on graph paper with 1×1 squares, then on graph paper with $\frac{1}{2} \times \frac{1}{2}$ squares, then $\frac{1}{4} \times \frac{1}{4}$ squares, and so on. The result looks like this:

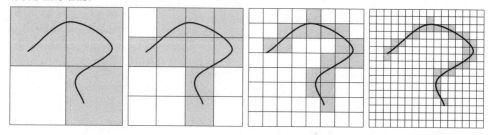

We've shaded the squares touched by the curve;

we want to count those boxes. Here are the results:

Grid size	1	1/2	1/4	1/8
Box count	3	9	18	33

Notice that when we halve the spacing between the grid lines, we roughly double the number of boxes needed to cover the curve. Here's why: Each box covers part of the length of the curve. When we cut the size of the boxes in half, we're going to need twice as many. We can express this as an equation like this:

$$N \propto \frac{1}{g} \qquad (A)$$

where g is the grid size and N is the number of boxes touched by the curve. The $\propto$ symbol stands for "is proportional to" and hides some of the inexactness of this relation. Had the curve been a simple, horizontal line segment, then we could have an exact equation. But because the curve twists around a bit, this relationship is quite good, but not perfect.

LET'S REPEAT THE BOX-COUNTING PROCESS with a two-dimensional figure: a disk of radius 1.

We repeatedly draw the disk on graph paper with box sizes 1×1, $\frac{1}{2} \times \frac{1}{2}$, $\frac{1}{4} \times \frac{1}{4}$, and so forth. Each time we shade the boxes touched by the disk. This includes the boxes that touch the bounding circle of the disk and all the boxes interior to the disk.

On 1×1 graph paper, if we place the center of the disk at a grid point, it's easy to see that it touches exactly four grid boxes. Here's the situation for finer grid sizes starting with $\frac{1}{2} \times \frac{1}{2}$.

For mathematicians, a *circle* is a one-dimensional curve and a *disk* is a two-dimensional figure consisting of a circle and the region interior to that circle.

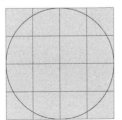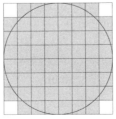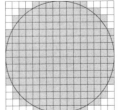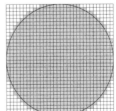

It's easy to see that for $\frac{1}{2} \times \frac{1}{2}$ graph paper, the disk touches 16 boxes. For $\frac{1}{4} \times \frac{1}{4}$ graph paper, the disk hits all but four of the 64 boxes drawn in the figure; the box count in this case is 60. For finer meshes the counting becomes tedious, but one can program a computer to do the counting for us. Here are the results:

Grid size	1	1/2	1/4	1/8	1/16	1/32	1/64	1/128
Box count	4	16	60	224	856	3332	13,104	51,940

A quick glance at these numbers shows that halving the grid size much more than doubles the box count. Here are the ratios:

$16 \div 4 = 4,$ $\quad 3332 \div 856 = 3.89,$

$60 \div 16 = 3.75,$ $\quad 13104 \div 3332 = 3.93,$ and

$224 \div 60 = 3.73,$ $\quad 51904 \div 13104 = 3.96.$

$856 \div 224 = 3.82,$

Roughly speaking, when we refine the grid size by a factor of two, the number of boxes touched increases by a factor of *four*. Furthermore, this "roughly speaking" becomes more accurate as the box size gets smaller. Let's see why.

When the grid spacing is small, the vast majority of the boxes touched by the disk lie completely inside the disk. There are plenty around the periphery, but their number pales by comparison to the huge number inside. When we cut the grid size in half, each interior box is subdivided into four smaller boxes, so that part of the count quadruples. For the boxes on the periphery, they too are cut into four by the finer mesh, but not all of the newly formed small boxes are touched by the disk so their number is not quite quadrupled.

By a similar logic, if we count the boxes touched by the disk in a grid and then make the grid spacing 10 times smaller, the number of boxes touched by the finer grid would be about 100 times more than the original. Every internal box on the first grid gets replaced by 100 tiny boxes in the refinement. The effect is muted for boxes on the boundary, but since they are much fewer in number than interior boxes, that effect is dwarfed in comparison.

We can express the relationship between the number of boxes and the grid size like this:

$$N \propto \frac{1}{g^2}. \qquad (B)$$

Here's another way to see why equation (B) is true. The area of a disk is given by the formula $A = \pi r^2$. In this example, we're considering a disk of radius 1, so that area is simply π.

We draw the disk on $g \times g$ graph paper and count the boxes that the disk touches; let's say the answer is N. Each of these N boxes has area g^2. The total area of these N boxes is about the same as the circle. So we have the relation

$$\text{area} = \pi \approx Ng^2.$$

Solving for N gives $N \approx \frac{\pi}{g^2}$. We can simplify this to $N \propto 1/g^2$, which is (B).

WE HAVE FOUND A QUANTIFIABLE WAY to distinguish one- and two-dimensional objects.

Relation (A) holds not only for the squiggle on page 194 but for any one-dimensional object. For a one-dimensional object, making the grid, say, ten times finer increases the number of boxes touched by the object by about a factor of 10.

Relation (B) holds not only for a disk, but for any two-dimensional figure. Refining a grid by a factor of 10 will increase the number of boxes touched by the object by about a factor of 100 because every interior box is replaced by 100 smaller boxes.

That is, we have the following:

Dimension	Box count formula
1	$N \propto 1/g^1$ (A)
2	$N \propto 1/g^2$ (B)

Notice that we attached the exponent 1 to g in formula (A). This doesn't change the formula but does set up some foreshadowing!

The dimension of Sierpinski's triangle

WE NOW HAVE A TOOL TO DISTINGUISH one-dimensional versus two-dimensional objects. We take the object, overlay it with successively finer grids, and count boxes. We either find that $N \propto 1/g^1$, signaling a one-dimensional object, or we find that $N \propto 1/g^2$ for a two-dimensional figure.

Let's see what happens when we apply this tool to Sierpinski's triangle. We place Sierpinski's triangle so it exactly fits inside a 1×1 square. The following pictures show the same figure embedded in grids of size $\frac{1}{2}$, $\frac{1}{4}$, $\frac{1}{8}$, and $\frac{1}{16}$:

Earlier in this chapter we constructed Sierpinski's triangle starting with an equilateral triangle. That doesn't fit so neatly in a 1×1 square: it's fine side-to-side, but it's a tad short vertically. The variant we consider here is constructed in exactly the same manner, but it starts with an isoceles triangle whose base and height have length 1.

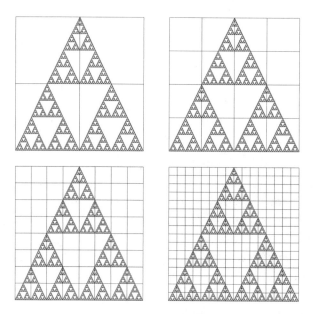

It's easy to see that all four boxes are touched in the $\frac{1}{2} \times \frac{1}{2}$ grid. In the $\frac{1}{4} \times \frac{1}{4}$ grid there are two untouched boxes in the upper left and two untouched boxes in the upper right, but all the rest of the boxes are touched by the figure. So for $g = \frac{1}{4}$ we have $N = 12$. Here's a chart for these and some finer grid sizes:

Grid size	1/2	1/4	1/8	1/16	1/32	1/64	1/128
Box count	4	12	36	108	324	972	2916

The question is: When we reduce the grid size by a factor of 2, does the box count double (like a one-dimensional object) or quadruple (like a two-dimensional object)?

Of course, the punch line is: Neither! Each box count in the chart is *exactly* triple the previous one. The box count for Sierpinski's triangle grows *faster* than one-dimensional objects but *slower* than two-dimensional objects. Its dimension sits between these two whole-number values.

WE CAN GIVE A PRECISE VALUE for the dimension of Sierpinski's triangle, but to do so requires some basic use of logarithms and a moderate amount of algebra. If that's uncomfortable, feel free to skip the next few paragraphs.

Our goal is to find a formula similar to equations (A) and (B)—something like this: $N \propto 1/g^d$. The exponent on d is the dimension.

When the grid size is $1/2^k$ (where k is a positive integer) we find that $N = \frac{4}{3} \cdot 3^k$. Here's the check:

k	1	2	3	4
$1/2^k$	1/2	1/4	1/8	1/16
$\frac{4}{3} \cdot 3^k$	4	12	36	108

Notice that the formula $N = \frac{4}{3} \cdot 3^k$ gives exactly the same values as presented in the earlier table.

Our problem is to find a number d so that $N \propto 1/g^d$. It turns out that taking logs of both sides is helpful:

$$N \propto \frac{1}{g^d} \Rightarrow \log N \sim -d \log g \Rightarrow d \sim -\frac{\log N}{\log g}.$$

The reason for the tripling comes from the self-similar nature of Sierpinski's triangle. Compare the middle two pictures (for grid sizes $\frac{1}{4}$ and $\frac{1}{8}$). In the left picture, Sierpinski's triangle touches exactly 12 boxes; that's easy to count. Now scan to the right, but don't try counting touched boxes. Instead, see this image as three small versions of the one on the left: lower left, lower right, top center. Each of these three portions contributes 12 touched boxes for a total of 36. Now scan to the farthest right image in the row. It contains three miniature copies of the previous one, giving a total of $3 \times 36 = 108$ touched boxes.

The symbol $\sim$ is a form of approximate equality whose accuracy gets better as the numbers involved get large.

We know that $N = \frac{4}{3} \cdot 3^k$ when $g = 2^{-k}$. Substituting these in the formula for d we have

$$d \sim -\frac{\log N}{\log g} = -\frac{\log\left(\frac{4}{3} \cdot 3^k\right)}{\log\left(2^{-k}\right)}$$

$$= \frac{\log \frac{4}{3} + k \log 3}{k \log 2} \sim \frac{\log 3}{\log 2} = \log_2 3,$$

which is approximately 1.5849625.

SIERPINSKI NOT ONLY HAS A TRIANGLE, he also has a carpet. Here are the steps to weave Sierpinski's carpet:

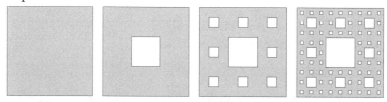

Taking this "to infinity" leads to an image like this:

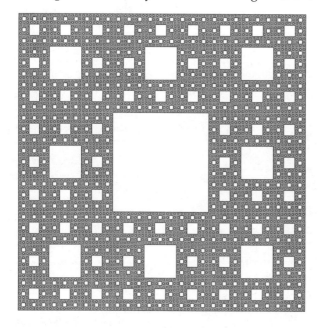

What is the dimension of this fractal? The answer is on page 203.

Pascal and Sierpinski

STUDENTS IN ALGEBRA CLASS TOIL AWAY multiplying polynomials, especially powers of $x + y$. Let's recall what these powers look like:

$$(x+y)^0 \longrightarrow 1$$
$$(x+y)^1 \longrightarrow x + y$$
$$(x+y)^2 \longrightarrow x^2 + 2xy + y^2$$
$$(x+y)^3 \longrightarrow x^3 + 3x^2y + 3xy^2 + y^3$$
$$(x+y)^4 \longrightarrow x^4 + 4x^3y + 6x^2y^2 + 4xy^3 + y^4$$
$$(x+y)^5 \longrightarrow x^5 + 5x^4y + 10x^3y^2 + 10x^2y^3 + 5xy^4 + y^5.$$

We can display the coefficients of these polynomials in a table that we call *Pascal's triangle*, which looks like this:

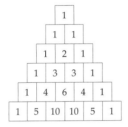

We've enclosed these numbers in squares because our next step is to color some of the squares black and some white. Specifically, we color black those squares containing an odd number and color white the squares containing an even number. The result looks like this:

Finally, do this for a lot of rows. Here's what we see when we extend this to the 64th row.

Isn't that simply wonderful?!

The Koch snowflake

WE CLOSE THIS CHAPTER ON FRACTALS with a beautiful shape created by Helge von Koch. Drawing Koch's snowflake depends on a single, simple step. Given a line segment, we draw an equilateral triangle whose base is the middle third of the original segment. Then we delete that middle portion of the original segment, leaving in its place the other two sides of the equilateral triangle. We now have a path consisting of four line segments, each one third the length of the original. And then the entire process is repeated; see the diagram in the margin.

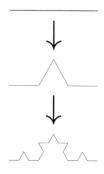

To get the full snowflake effect, we begin with an equilateral triangle (as opposed to a single line segment). We perform the basic Koch step to each side of the triangle, always pointing the bump outward. And then we repeat. And repeat. Forever.

Here's what the process looks like:

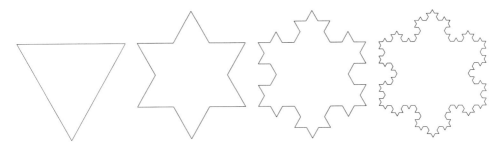

We continue this process "to infinity" and the result is Koch's snowflake.

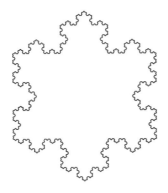

The dimension of Sierpinski's carpet. In this case, using graph paper with mesh sizes 1, 1/3, 1/9, 1/27, and so forth works nicely. We get a chart that looks like this:

Grid size	1	1/3	1/9	1/27	1/81
Box count	1	8	64	512	4096

Each time g is refined by a factor of 3, the box count N is multiplied by 8. We can capture this in a formula:

$$N = 8^{k-1} \quad \text{when } g = 1/3^k.$$

Using the expression $d \sim -(\log N)/(\log g)$ (see page 200) we get

$$d \sim -\frac{\log\left(8^{k-1}\right)}{\log\left(3^{-k}\right)} = \frac{(k-1)\log 8}{k \log 3} \sim \frac{\log 8}{\log 3}.$$

Thus the dimension of Sierpinski's carpet is $\log_3 8 = 1.89278926$.

18
Hyperbolic Geometry

Euclid's Postulates

MATHEMATICIANS ARE DEFINITION FANATICS. We demand that every concept be specified by a crystal-clear, unambiguous definition. To do this, each mathematical idea is precisely built from simpler ideas. Triangles are built from line segments. Rational numbers are specified as ratios of integers.

The tower of mathematical definitions has to begin somewhere. For the Greeks, the foundation was geometry.

Modern mathematicians rest this edifice on the concept of *sets*—unordered collections of things.

Euclid did not attempt to define the basic concepts of points, lines, and planes. Rather, he took a different approach: He *assumed* certain fundamental properties that these concepts possess. Such assumptions are called *postulates* or *axioms*.

To kick-start geometry, Euclid proposed five basic postulates. Roughly translated, they say the following:

1. Given two points, there is a unique line that contains those two points.

2. Given a line segment, that line segment can be extended into a line.

3. Given a point and a distance, there is a unique circle whose center is that point and whose radius

is that distance.

4. Any two right angles are of equal measure.

5. If two lines meet a given line, and if the resulting interior angles on the same side of the given line are less than right angles, then the two lines must intersect, as illustrated here:

Euclid gives a definition of *right angle*: it's one of the angles formed when two lines cross and the four angles created are of equal measure.

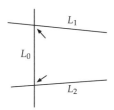

The first four postulates are simple and not difficult to understand. The fifth, however, is quite a mess. Let's examine what it says.

Call the given line L_0 and the two lines that intersect L_0 we call L_1 and L_2. Lines L_0 and L_1 form angles, as do lines L_0 and L_2. The figure in the margin illustrates this.

The postulate asks us to consider the situation in which the interior angles (on the same side of L_0) are less than right angles. The arrows in the diagram point to the angles that Euclid had in mind. They are both on the same side of L_0 and they are *interior* in the sense that they inwardly face each other.

We now come to the crux of the postulate. If those two angles are both less than right angles, then lines L_1 and L_2 must intersect. They are not shown intersecting in the picture, but it's not hard to imagine that, when extended, they eventually must cross.

With these five postulates taken as given, Euclid proceeds to prove a host of lovely theorems.

EUCLID'S FIFTH POSTULATE IS UGLY. It ugliness stands in contrast to the simple elegance of the first four. Mathematics is an aesthetic art as much as it is a practical problem-solving tool, so cleaning up

Euclid's foundation is appealing.

One approach is to replace Euclid's fifth postulate with something simpler. We offer this substitution:

5′. Given a line and point not on that line, there is a unique line through the point that does not intersect the original line.

This alternative version of Euclid's fifth postulate is known as the *Parallel Postulate*. Let's see what it says.

We are given a line L and a point P that is not a point of L. See the diagram. Postulate 5′ asserts that there is another line (shown dashed) through P that does not intersect L. Furthermore, there is only one such line (hence the word *unique* in the statement of the postulate).

Mathematicians have shown that Euclid's fifth postulate and the Parallel Postulate are logically equivalent. That means the theorems we can prove using the first four postulates together with postulate 5 are exactly the same as the theorems we can prove using the first four together with 5′.

While the statement of 5′ is a bit simpler than 5, it's still not as crisp and elegant as the first four. Can we get rid of it? In other words, might it be possible to prove the Parallel Postulate as a theorem, and not take it as a foundational assumption?

GIVEN A LINE L AND A POINT P NOT ON L the Parallel Postulate makes two claims: First, there is a line through P that is parallel to (meaning: does not intersect) L. Second, there can't be more than one line through P that doesn't intersect L.

A natural way to attack this problem is to try *proof by contradiction*—this is a method we introduced in Chapter 1. Here's how the logic would work.

(A) To prove the *existence* of a line through P parallel to L we *suppose* there is no line through P that is

> Lines in a plane that don't intersect are called *parallel*, and so this statement is known as the *Parallel Postulate*.
>
> We used the Parallel Postulate to prove that the sum of the angles of a triangle is 180°; see page 143.

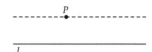

parallel to L.

(B) To prove the *uniqueness* of such a line, we *suppose* that there are two (or more) lines through P that are parallel to L.

In either case, we proceed with logical reasoning until we arrive at a contradiction. The contradiction signals a fundamental error, namely (A) or (B)—whichever we presumed to be true:

- If supposing there is no parallel leads to a contradiction, then there must be a parallel.

- If supposing there are two (or more) parallels leads to a contradiction, then there can be at most one.

This is the approach mathematicians tried—and failed. To be sure, things got *weird* (triangles in which the angle sum is not 180°) but no contradiction was found.

That's OK. Mathematicians do not imagine they will be able to solve every problem they face. We keep slogging away and hand off unfinished business to the next generation, hoping our successors will have better ideas than we had.

In the case of the Parallel Postulate, better ideas did emerge, but not in the way one might have expected.

What is a line?

A LINE IS A SET OF POINTS, as is a circle or a triangle. Only certain sets of points are deemed to be *lines*.

Intuitively, we understand what a line is: it's thin (has no width), straight, and continues forever in two opposing directions. But this description is not a complete mathematical definition. What does *straight* mean? That's a difficult concept to pin down.

Most mathematicians hoped that replacing the Parallel Postulate would lead to a contradiction. Nikolai Lobachevsky, a nineteenth-century Russian mathematician, is credited with realizing that (B) does not lead us to a contradiction, but rather to a new type of geometry, often called Lobachevskian geometry in his honor.

So, as we mentioned, Euclid took a different approach; here is how we think about points and lines today. We have things called *points* and certain collections (sets) of points we call *lines*. If these things (the things we are calling points and the things we are calling lines) satisfy Euclid's postulates, then we have a system we call *Euclidean geometry*.

If we change Euclid's assumptions about the fundamental properties of points and lines, then we arrive at a different type of geometry. Let's look at a simple example. We keep Euclid's first postulate, that we repeat here:

1. Given two points, there is a unique line containing those points.

And we include a second postulate that reverses the roles of points and lines:

1'. Given two lines, there is a unique point contained in both of those lines.

Notice that Postulate 1' implies that there are no parallel lines; it asserts that every pair of lines intersect. This is not the case in Euclidean geometry.

Let's see that suitably chosen "points" and "lines" satisfy conditions 1 and 1'. For this example, there are exactly seven points that we unimaginatively name 1, 2, 3, 4, 5, 6, and 7. And there are exactly seven lines; here they are:

$$\{1,2,3\} \quad \{1,5,6\} \quad \{1,4,7\} \quad \{2,5,7\}$$
$$\{2,4,6\} \quad \{3,4,5\} \quad \{3,6,7\}.$$

This system of seven points and seven lines is called the *Fano plane*.

These seven "lines" are nothing like Euclidean lines. Every line has only three points!

With a bit of care, we can check that this system of seven points and seven lines satisfies postulates 1 and 1'.

- Check postulate 1. Pick two points, any two points: let's say 2 and 5. These two points are both in the line $\{2,5,7\}$, and there is no other line that contains them both.

> Here is a way to define *Euclidean* points and lines. A *point* is a pair of real numbers (x, y). A *line* is a set of points (x, y) that satisfy an equation of the form $ax + by + c = 0$ where a and b are not both zero. With these definitions in place (and suitable definitions of circle and angle), one can *prove* that Euclid's postulates are satisfied.
>
> Representing points as pairs of numbers and lines as solutions to equations sets up the *Cartesian* plane, named after the mathematician and philosopher René Descartes.

If you check all possible pairs of numbers, you'll find there is always a line—and only one line— that includes those two numbers.

- Check postulate 1'. Pick two lines, any two lines: let's say $\{1, 4, 7\}$ and $\{3, 4, 5\}$. These lines both contain point 3 and that's the only point they have in common.

If you check all possible pairs of lines, you'll find that they always include a common point—and only one common point.

It's strange to speak of geometry without a picture. Luckily, there's a nice way to represent this system with a diagram as shown in the figure. The seven points are represented as dots and the seven lines are indicated by segments (for most of the lines) and by a circle (for the line $\{2, 4, 6\}$).

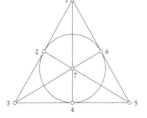

The point (hah!) is that we have gathered up a bunch of things we call *points* and then designated certain collections (sets) of them that we call *lines*. If these so-called points and lines satisfy conditions that make geometric sense, then the names *point* and *line* are reasonable, even if they are nothing like the points and lines Euclid had in mind.

An entire plane inside a disk

WE HAVE BECOME RATHER PERMISSIVE is using the words *point* and *line*. We may choose anything as points and collections of those points as lines, so long as they satisfy appropriate postulates. What is *appropriate*? To Euclid, the relevant assumptions are his five postulates that we presented at the beginning of this chapter.

With this context, we present another collection of "points" and "lines" to create *hyperbolic geometry*. The points of this geometry are all the points lying inside a fixed circle. The entire region interior to this circle we call the *hyperbolic plane*.

> This model of the hyperbolic plane is known as the Poincaré disk model, in honor of the nineteenth-century French mathematician, Henri Poincaré.

The lines of the hyperbolic plane are certain circular arcs. This is confusing: How can an arc be a line? Arcs are curved and lines are straight! To make this easier to swallow, we use the expression *hyperbolic line* to distinguish it from its unbent cousin.

The following are the hyperbolic lines in the hyperbolic plane.

- Draw a circle that meets the hyperbolic plane in two right angles (see the arrows in the diagram on the facing page). The portion of that circle that's interior to the hyperbolic plane is a hyperbolic line.

- Draw a line through the origin of the hyperbolic plane. The portion of that line that lies interior to the hyperbolic plane is also a hyperbolic line.

The figure on the next page shows three hyperbolic lines drawn in the hyperbolic plane.

The hyperbolic plane is the interior of the finely dotted circle. In this figure we see two hyperbolic lines that are circular arcs that are perpendicular to the boundary and one hyperbolic line that is a diameter line segment. Note that the end points of these arcs and of the diameter are not part of the

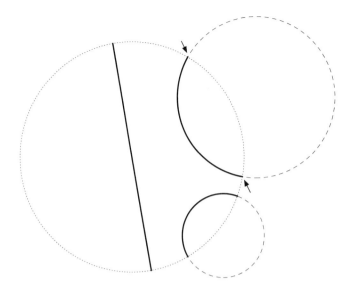

Three lines in the hyperbolic plane.

respective hyperbolic lines. (The coarsely dashed parts of the circles are not part of the hyperbolic lines; they are there just to show how the arcs are parts of circles that meet the finely dotted circle at right angles.)

For the figure in the margin we present three lines. Two of the lines cross each other and the third is parallel to them both! Such an arrangement of lines is impossible in the Euclidean plane.

Implications

THINGS ARE RATHER DIFFERENT IN THE HYPER-BOLIC PLANE. Many geometry "facts" about the Euclidean plane don't hold in the case of the hyperbolic plane.

To begin, triangles are rather different. In the Euclidean plane, the sum of the angles of a triangle is exactly 180°. In the hyperbolic plane, the sum is always less than 180. (In Chapter 13 we proved that

the sum of the angles of a triangle must equal 180, but that proof used the Parallel Postulate as a key step.)

In the Euclidean plane, the area of a triangle can be as big as we like. In the hyperbolic plane, there's a maximum area for all triangles and there's a simple formula to calculate the area. If the three angles of a triangle add up to s, then the area of the triangle is given by $K(180 - s)$ where K is a specific number. This formula implies that if two different triangles have the same angles, then they have the same area. In Euclidean geometry this isn't so: two triangles with, say, angles 35-65-80 have the same shape (we say they are *similar*) but may be rather different sizes. In the hyperbolic plane, two 35-65-80 triangles not only have the same area, they are actually congruent!

Technicality: Rescaling the Euclidean plane doesn't actually change it. Hyperbolic planes, however, are a bit different when rescaled. The number K in the triangle area formula depends on that scale.

A RECTANGLE IS A FOUR-SIDED SHAPE (quadrilateral) in which all four angles are 90°. Here's an interesting fact about rectangles in the hyperbolic plane: *There are no rectangles in the hyperbolic plane!* The figure in the margin represents a quadrilateral in the hyperbolic plane that has three right angles, but the fourth angle is less than 90°.

Why are rectangles impossible? Think about a quadrilateral R in the hyperbolic plane. Cut R in two by joining a pair of opposite corners by a line segment, giving two triangles. The sum of the angles in each of those two triangle is less than 180, so when we reassemble them into the quadrilateral R, the total of all angles would be less than 360. That means that R's four angles cannot all be 90°.

THE EUCLIDEAN PLANE CAN BE TILED by equilateral triangles, squares, and regular hexagons. It's not possible to tile the plane with regular pentagons. Here's why: The corners of a regular pentagon are 108° angles. If we join three regular pentagons at cor-

A tiling of the plane using polygons is called a *tessellation*. Here we consider tessellations in which all the tiles are regular n-gons.

ners, the angles only add up to 324° and hence leave a gap. But we can't squeeze four of them together as those angles add up to 432°, which exceeds the exact amount required: 360°.

However, the angles of a regular n-gon in the hyperbolic plane do not depend just on n. That means, for example, we can make a regular pentagon all of whose angles are 90° (see the figure in the margin). We can take four such pentagons at a corner of each and exactly fill the 360° space. Continuing this way, we can produce an amazing tiling of the hyperbolic plane, as shown in the picture.

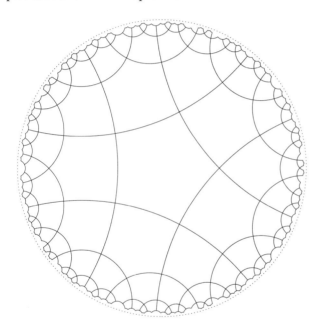

All the pentagons in this picture are exactly the same size and shape. The fact that they look smaller as they near the dotted boundary is simply an artifact of how we make a picture of the hyperbolic plane. The tiles in this picture are identical to each other. They are all regular pentagons whose angles are all 90°; that's why they fit together perfectly.

For your viewing pleasure, here are two more

The tessellations of the hyperbolic plane have inspired many artists, most notably M. C. Escher. We recommend hyperbolictessellations.com for many lovely examples.

tessellations of the hyperbolic plane.

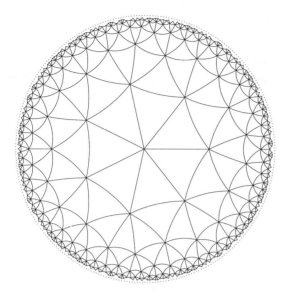

Triangles meeting 7 at a corner.

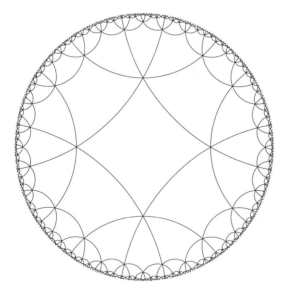

Quadrilaterals meeting 6 at a corner.

PART III: UNCERTAINTY

19
Nontransitive Dice

THE WORLD IS OBSESSED WITH RANKINGS. We rank athletes, sports teams, hospitals, restaurants, movies, pop songs, students, colleges, cities, jobs, cars, and on and on. We love to know which is "the best" and which are in the all-important "top ten."

This is a bunch of nonsense: fun nonsense, but nonsense nonetheless. In some cases, the nonsense derives from the subjectivity of the ranking methodology. If a certain restaurant in your town is deemed to be "the best" that does not mean it's your favorite. Your preferences might not match the assessment of the critics who rate restaurants, and their assessments might contradict each other.

It's possible to choose an objective ranking system and still get nonsensical results. We can judge movies based on how much money they earn; that's objective and measurable. One can reasonably argue that the better a film is, the more people will be willing to pay to see it. But a top-grossing film might bore you while an unknown, low-budget indy film might delight you. A film's sales might have more to do with marketing than with quality.

But suppose we strip away all subjectivity and we are in universal agreement on how we compare the competitors. Let's boil this ranking idea down to its mathematical core. Does the nonsense vanish?

A game of two dice

LET'S PLAY A SIMPLE GAME. We each take a die, roll, and whoever's die shows the larger number wins. If we're playing with two standard dice—both are marked with the values from one to six—then it doesn't make sense to say that one die is better than the other. They're equally matched.

So let's change the spots on the two dice. We name the dice A and B and mark their six sides with the values as shown in the figure.

A | 2 | 3 | 4 | 15 | 16 | 17 |

B | 5 | 6 | 7 | 8 | 9 | 18 |

Which die is better, A or B? If you were playing this game, which would you prefer?

To answer, we consider all the ways the dice might land. If die A shows 2, that 2 might be matched with any of the six values on B. If A shows a 3, again, there are six values on the faces of B with which it might be matched. All told there are $6+6+6+6+6+6 = 6 \times 6 = 36$ possible outcomes, and all 36 are equally likely. Sometimes A wins and sometimes B wins (since they have no number in common, there's never a tie). Which wins more often?

Let's make a chart showing all 36 possible combinations and determine in each case which die wins: A or B. Here's the chart:

		\multicolumn{6}{c}{A}
		2	3	4	15	16	17
	5	B	B	B	A	A	A
	6	B	B	B	A	A	A
	7	B	B	B	A	A	A
B	8	B	B	B	A	A	A
	9	B	B	B	A	A	A
	18	B	B	B	B	B	B

It's now easy to see that B is the better die. In head-to-head competition, B beats A more often than not. Counting the number of As and Bs in the chart, we see that A wins on average 15 times out of 36, but B wins 21 times out of 36.

Gamblers would say A's odds of winning are 15 : 21 and that B's odds are 21 : 15. Mathematicians express this as probabilities: The probability A wins is 15/36 (or about 42%) and the probability that B wins is 21/36 (or about 58%).

Any way you like to say it: B is better than A.

A challenger

NOW WE THROW A THIRD DIE INTO THE MIX. Enter the challenger! Let C be the die whose sides are marked with six numbers as shown.

C | 1 | 10 | 11 | 12 | 13 | 14 |

With gusto, C challenges B to the same game. Both dice are thrown and whichever presents the larger value wins. Which die is better, B or C? As before, we present a chart showing all 36 possible outcomes and check to see which die is more likely to be victorious.

		C					
		1	10	11	12	13	14
	5	B	C	C	C	C	C
	6	B	C	C	C	C	C
B	7	B	C	C	C	C	C
	8	B	C	C	C	C	C
	9	B	C	C	C	C	C
	18	B	B	B	B	B	B

Notice that C is much more likely to win than B. The probability that C triumphs is 25/36 (about 69%) while B only wins with probability 11/36 (about 31%).

In head-to-head competition, C is better than B, and B is better than A.

This makes C the "best" of the three, right?

Triumph of the underdog

It seems that of the three dice, A is the weakest and C is the best. What happens when A competes against C? Surely C is superior to A.

As before, we construct a chart of all the possible outcomes like this:

	A					
C	2	3	4	15	16	17
1	A	A	A	A	A	A
10	C	C	C	A	A	A
11	C	C	C	A	A	A
12	C	C	C	A	A	A
13	C	C	C	A	A	A
14	C	C	C	A	A	A

Behold! A is better than C. Die A wins with probability 21/36 (about 58%) but C only wins with probability 15/36 (about 42%).

We reach the following surprising trio of conclusions:

- B is better than A,
- C is better than B, and
- A is better than C.

There is no sense in which one of these dice is the "best" and to attempt to rank them is utter nonsense.

How many other rankings that we encounter in daily life are also nonsensical?

Further examples

HERE'S ANOTHER SET OF DICE for you to analyze; this example is the invention of Bradley Efron, a statistics professor at Stanford.

#1 | 1 | 1 | 5 | 5 | 5 | 5
#2 | 4 | 4 | 4 | 4 | 4 | 4
#3 | 3 | 3 | 3 | 3 | 7 | 7
#4 | 2 | 2 | 2 | 6 | 6 | 6

Consider the four dice shown in the figure. Work out the probability that 1 beats 2, that 2 beats 3, that 3 beats 4, and that 4 beats 1. Which die is better in each of these matchups? How would you rank these dice?

The answers are on the following page.

DO YOU PLAY POKER? In particular, do you play Texas Hold 'Em? Imagine two people are playing and you're allowed to peek at their pocket (hidden "hole") cards. Let's say the first player holds A♠ K♡ and the second player holds 10♢ 9♢. Who has the better chance of winning? The first player has higher value cards, but the second player has a possible straight and a possible flush started.

To figure out which is better, we have to consider the five, face-up community cards that will be dealt. There are 48 cards unaccounted for (the 52 in the deck minus the 4 in the two players' hands). In principle, we could go through all the possible ways the community cards might be drawn, and for each set of five community cards, figure out which player—the one holding A♠ K♡ or the one holding 10♢ 9♢— wins (or if the result is a tie). There are nearly two million possible ways the community cards can be selected. That's far too many to do the calculation by hand, but it's done easily by computer. A quick internet search for `poker calculator` leads to various web sites that can do this calculation speedily.

The exact number of ways to select 5 cards from the remaining 48 is $\binom{48}{5}$ which equals 1712304.

Using such a calculator we learn that A♠ K♡ wins 58.6% of the time, 10♢ 9♢ wins 41% of the time, and they tie 0.4% of the time.

Conclusion: A♠ K♡ is a better holding than 10♢ 9♢.

Now it's your turn. Find a poker odds calculator and use it to compare these two additional matchups:

- 10♢ 9♢ versus 2♣ 2♡.

- 2♣ 2♡ versus A♠ K♡.

Use the results that the calculator found to rank the three selections of pocket cards: A♠ K♡, 10♢ 9♢, and 2♣ 2♡.

The answers are below.

Efron's dice: Building tables for the four matchups reveals that the probability the better die wins is 2/3 in all four cases: #1 is better than #2, #2 is better than #3, #3 is better than #4, and #4 is better than #1. Ranking these four dice is absurd.

Texas Hold 'Em: Using a poker calculator we arrive at the following results:

- A♠ K♡ beats 10♢ 9♢ 58.6% of the time.
- 10♢ 9♢ beats 2♣ 2♡ 52.9% of the time.
- 2♣ 2♡ beats A♠ K♡ 52.3% of the time.

This implies that A♠ K♡ is better than 10♢ 9♢, that 10♢ 9♢ is better than 2♣ 2♡, and that 2♣ 2♡ is better than A♠ K♡. Ranking these possible pocket cards is nonsense.

All this assumes that only two people are playing Texas Hold 'Em. What happens if there are three players at the table and all three pocket card hands are in play at once?

20
Medical Probability

A NEW DIAGNOSTIC TEST for a rare disease has been announced. It's a decently reliable test and you decide to have yourself tested and, much to your dismay, the test comes back positive. How worried should you be?

It's difficult to quantify worry, and it would be natural for anyone in this situation to be concerned, so let's modify the question: How likely is it that you have this rare disease?

To answer, we need some idea of how reliable the test is and, as we shall see, how rare the disease is. So here's some data.

The rare disease affects 0.1% of the population: one person out of every thousand has this worrisome condition.

The test for this disease is not perfect—no diagnostic test is. Let's suppose it is 98% reliable. By this we mean the following:

- Given 100 healthy people, the test will correctly report that 98 are fine but erroneously report that 2 are sick.

- Given 100 sick people, the test will correctly report that 98 are ill but falsely report that 2 are healthy.

Of course, we'd like a more reliable test, but let's imagine this is the only diagnostic available.

Our question is this: Your test comes back positive. What is the probability you have the disease?

The answer seems obvious. We just explained that the test is 98% reliable. It would appear that the probability you have the disease is 98%. Right?

IMAGINE A CITY OF ONE MILLION PEOPLE. In this population of one million people, one in a thousand has the disease. That means there are 1000 sick people and the other 999,000 are healthy.

We now give these million people the diagnostic test. Let's count how many positive results we see given that the test is 98% effective.

- Of the one thousand sick people, most, but not all, will get positive results. The number is $1000 \times 0.98 = 980$.

- Of the 999,000 healthy people, most will get the good news that they do not have the disease, but 2% will get false positive results. That's $999,000 \times 0.02 = 19,980$ positive results.

All told, there are $980 + 19,980 = 20,960$ people for whom the test results are positive.

And now we can answer the question: What is the probability you have the disease given that your test result came back positive?

Of the more than twenty thousand people who get a positive test result, fewer than one thousand are, in fact, ill. The precise probability is

$$\frac{980}{20,960} = 4.7\%.$$

The probability you have the disease is *not* 98%. In fact, the chance that you are ill is under 5%!

DOES THIS MEAN THE TEST IS WORTHLESS? Not at all.

First, if your doctor has a reason to suspect that you have the disease, then you are no longer a "random" patient. If you have certain symptoms, the odds that you have the disease are not one in a thousand, but might be—let's say—one in four. In this case, a positive test result is much more meaningful than if it were administered without cause.

Second, if this disease is dangerous, using this 98%-accurate test might be a good way to screen a large population. Those who receive positive results can then be given a second—presumably more costly—follow-up test to hone in on an accurate diagnosis.

Of course, getting a negative result can be reassuring. But how reassured should you feel? (Solution on page 229.)

Figure for yourself: Suppose you are in a subpopulation in which 25% of the patients have the disease. If you get a positive test result, what is the probability you have the disease? Answer on page 229.

IT'S CERTAINLY COUNTERINTUITIVE that a test that is 98% accurate can be so misleading, but the calculations speak for themselves. However, looking at raw numbers doesn't help our intuition. What helps is to draw a picture.

In the figure on the next page, the large rectangle represents an entire population. Contained in that population is a small region (in the upper left) that represents the sick people, and the rest of the region represents the healthy individuals. The gray stripe across the top of the diagram represents those people (from both groups) who receive a positive test result. The white areas are those people who receive a negative test result (again, in both groups).

The diagram illustrates the key features of this problem:

- The disease is rare—only a small fraction of the entire population falls into the "sick" category;

- The test correctly diagnoses most of the sick people—the gray bar covers nearly all of the "sick" region; and

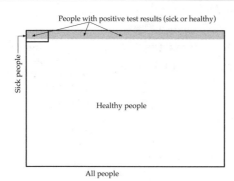

Note: The proportions in this diagram do not exactly match the percentages given (0.1% sick, 98% test accuracy).

The two regions represent those who are sick (small box on the upper left) and those who are healthy (the rest of the full rectangle). The gray bar, running horizontally through both regions, represents those who receive positive test results. Notice that most people are diagnosed correctly. Nearly everyone in the sick portion receives a positive test result (gray) and nearly everyone in the healthy region receives a negative result (white).

- The test correctly diagnoses most of the healthy people—the gray portion is just a small part of the "healthy" region.

The key point is that *most* of the gray stripe comprises healthy people, so getting a positive test result is more likely to mean that you received one of the false positive results rather than that you are sick.

Conditional probability*

WE'VE CALCULATED THE PROBABILITY that a patient has the disease under the assumption of a positive test result. We did this by imagining a hypothetical

*This discussion is meant for those who have studied probability and just need a bit of refresher. Other readers can safely skip this section.

population of one million people and calculated various subpopulations to derive our answer. A less *ad hoc* approach would be to use the language and notation of probability directly, and we close this chapter with an explanation of how this is done.

For an event A, we write $P(A)$ to denote the probability that A occurs. We write $P(\overline{A})$ for the probability that A does *not* occur and we have $P(\overline{A}) = 1 - P(A)$.

For events A and B, we write $P(A \wedge B)$ for the probability that *both* A and B occur.

The notation $P(A \mid B)$ represents the probability that A occurs given that B occurs; this is the *conditional probability* of A given B. Bayes' formula tells us that

$$P(A \mid B) = \frac{P(A \wedge B)}{P(B)}.$$

The assumptions given to us for this diagnostic testing problem can be expressed as follows. Let S be the event that a person is sick with the disease and let T denote receiving a positive test result. We are given the following:

- The disease affects 0.1% of the population, so $P(S) = 0.001$.

- The test correctly reports that a person has the disease 98% of the time, so $P(T \mid S) = 0.98$.

- The test correctly reports that a person is healthy 98% of the time, so $P(\overline{T} \mid \overline{S}) = 0.98$. Alternatively expressed, the test gives a false positive 2% of the time: $P(T \mid \overline{S}) = 0.02$.

The question is: What is the probability a patient is sick given a positive test result?

In symbols, we want to determine $P(S \mid T)$. We know that this is equal to $P(S \wedge T)/P(T)$. So we need to work out $P(S \wedge T)$ and $P(T)$.

Let's start with $P(S \wedge T)$, which is the same as $P(T \wedge S)$. By Bayes' formula we have

$$P(T \mid S) = \frac{P(T \wedge S)}{P(S)},$$

and we know that $P(T \mid S) = 0.98$ and that $P(S) = 0.001$. Therefore

$$P(S \wedge T) = P(T \wedge S) = P(T \mid S)P(S) = 0.98 \times 0.001 = 0.00098.$$

Next we calculate $P(T)$. We know that $P(T \mid S) = 0.98$ and that $P(T \mid \overline{S}) = 0.02$. We can write $P(T)$ as $P(T \wedge S) + P(T \wedge \overline{S})$. Observe:

$$P(T \wedge S) = P(T \mid S)P(S) = 0.98 \times 0.001 = 0.00098$$
$$P(T \wedge \overline{S}) = P(T \mid \overline{S})P(\overline{S}) = 0.02 \times 0.999 = 0.01998$$
$$\Rightarrow \quad P(T) = P(T \wedge S) + P(T \wedge \overline{S})$$
$$= 0.00098 + 0.01998 = 0.02096.$$

Finally we substitute into Bayes' formula one more time:

$$P(S \mid T) = \frac{P(S \wedge T)}{P(T)} = \frac{0.00098}{0.02096} \approx 0.0468,$$

confirming our earlier calculation.

If you have symptoms: Suppose, because of your symptoms, your chances of having the disease are 25%. If you get a positive result, what is the likelihood you are ill? Imagine a population of one million people in your situation. Of these, 250 thousand are ill and the remaining 750 thousand are healthy.

- Of the 250 thousand with the disease $250{,}000 \times 0.98 = 245{,}000$ will get positive test results.
- Of the 750 thousand disease-free people $750{,}000 \times 0.02 = 15{,}000$ will get false positive results.

All together, 260,000 people that get positive results, and 245,000 are, indeed, ill. Thus the probability you are ill is $245 \div 260 = 94.2\%$.

If you get a negative result: Suppose you take the test and the result is negative. What is the probability you are, indeed, healthy?

In our city of one million people, one thousand have the disease and 999 thousand do not. How many get negative results?

- Among the one thousand sick people, 2% will get false negative results. That's $1000 \times 0.02 = 20$ people.
- Among the 999 thousand healthy people, 98% will get negative results. That's $999{,}000 \times 0.98 = 979{,}020$ people.

So the probability you are healthy is

$$\frac{979{,}020}{20 + 979{,}020} = \frac{979{,}020}{979{,}040} = 99.998\%.$$

This might seem like fantastic news, but remember: Even without the test, the probability that you are healthy is 99.9%. So the value added by the test is miniscule.

21
Chaos

WHAT MAKES AN EVENT UNPREDICTABLE? The previous chapters deal with ideas from *probability*. The core idea from probability theory is that some phenomena are *random*: they cannot be exactly predicted because they are *indeterminate*. It is reasonable, and highly effective, to model various real-world phenomena, such as the roll of a die, as random.

But is the roll of a die truly random? Perhaps if we knew every detail about the toss of a die—from its rotational speed to the air currents in the room to the coefficient of friction of the surface on which it lands—we could, with certainty, determine which side of the die would land face up. Perhaps the roll of a die isn't random; it's just difficult.

Is anything random? Physicists tell us that some physical phenomena are truly random; this is a central tenet of quantum mechanics. The behavior of tiny particles such as electrons and photons cannot be predicted because randomness is fundamental to how they operate.

Other physical, biological, and social phenomena are exceedingly well modeled by probability theory, and that's terrific. But are they random? Perhaps they're just terribly complicated.

This brings us to the question for this chapter: Can a system be really simple, completely deterministic, but utterly unpredictable?

Functions

THE UNIFYING IDEA IN THIS CHAPTER IS FUNCTION ITERATION. By *iteration* we simply mean repeatedly doing the same thing over and over again. Let's review what mathematicians mean by *function*.

Functions can be thought of as "black boxes" that transform one number into another. We imagine this box has an input slot into which we insert numbers, a crank to turn to make the box do its work, and an output chute from which the result emerges.

We are only considering functions that transform numbers to numbers. In general, functions convert all sorts of mathematical objects into other mathematical objects.

For example, imagine a box that does the following operations. It takes the number it receives, squares it, adds one to the result, and then spits out the result. Let's give this function a name; we call it the square-it-and-add-one function. Here's a picture of it working on the number 3:

$$3 \longrightarrow \boxed{\text{square-it-and-add-one}} \longrightarrow 10$$

Describing the action of a function in words is cumbersome; it's clearer to use mathematical symbols to indicate what's happening. For the number 3, we first square (so 3 becomes $3^2 = 3 \times 3$) and then add one to give $3^2 + 1$, which equals 10. What is the result when we insert the number 4 into this function? We get $4^2 + 1 = 17$.

Rather than using long names (such as "square-it-and-add-one"), mathematicians tend to use single letters to name functions, and most often the letter we choose is f. To show the action of f on a number, we place that number in parentheses immediately after the name of the function, like this: $f(4)$.

With this notation, we have a handy way to write a rule that precisely defines the function. We write:

$$f(x) = x^2 + 1.$$

This shows that for a given number, x, the result of applying the function f to that number is $x^2 + 1$.

Here's another example. Define a new function g by
$$g(x) = 1 + x + x^2.$$
What is $g(3)$? We do this calculation by substituting 3 for x in the definition of g and we get
$$g(3) = 1 + 3 + 3^2 = 1 + 3 + 9 = 13.$$

Functions can be combined by performing one operation after another. Let's tease apart what $f(g(2))$ means (using the same f and g we just defined).

The expression asks us to compute f of something; of what? We compute f of the value $g(2)$. What is $g(2)$? By the definition of g it is $1 + 2 + 2^2 = 1 + 2 + 4 = 7$. We now calculate f of that: $f(7) = 7^2 + 1 = 50$. Putting this all on one line we have this:
$$f(g(2)) = f(7) = 50.$$

Now please check your understanding of this by calculating $g(f(2))$. The answer is *not* 50. You can find the answer on page 244.

LET'S RETURN TO THE NOTION OF ITERATION. As previously mentioned, *iteration* simply means repeating the same thing over and over again. To reiterate: *iteration* means to do the same thing over and ... (OK, I hope you get the joke).

Let's think about the function $f(x) = x^2 + 1$. The notation $f(f(x))$ means we do the operation f twice: We take the number x, apply the function f to get a result, and then apply f to that result to get the final answer. Here's an example:
$$f(f(2)) = f(2^2 + 1) = f(5) = 5^2 + 1 = 26.$$
We can iterate more than twice. Here we do it thrice:
$$f(f(f(2))) = f(f(5)) = f(26) = 26^2 + 1 = 677.$$

As we get past three repetitions of the function, it is difficult to read the notation. So instead of writing $f(f(f(f(x))))$, we write $f^4(x)$ where we understand the exponent does not mean repeated multiplication but rather repeated function application. For a positive integer n, the symbol $f^n(x)$ means this:

$$f^n(x) = \underbrace{f(f(f(\ldots f(x)\ldots)))}_{n \text{ times}}.$$

Iterating the logistic map

IN THIS SECTION WE CALCULATE THE ITERATES of functions of the form $f(x) = mx(1-x)$ where m is a specific number. This family of functions are known as *logistic maps*. In all cases, we begin with the value $x = 0.1$, iterate the function, and see what happens. We begin with this function:

The word *map* is a synonym for function.

$$f(x) = 2.5x(1-x).$$

Starting with $x = 0.1$, we first calculate

$$f(0.1) = 2.5 \times 0.1 \times (1 - 0.1) = 2.5 \times 0.1 \times 0.9 = 0.225.$$

We apply f again:

$$\begin{aligned} f^2(0.1) = f(0.225) &= 2.5 \times 0.225 \times (1 - 0.225) \\ &= 2.5 \times 0.225 \times 0.775 = 0.4359375. \end{aligned}$$

Let's get some computer help. Writing a program to calculate the iterates of f we have the following result:

n	$f^n(0.1)$
1	0.225
2	0.4359375
3	0.614739990234
4	0.592086836603
5	0.603800036311
6	0.598063881154
7	0.600958688032
8	0.599518358277
9	0.600240240915
10	0.599879735253

Notice that the successive iterations get closer and closer to 0.6. There's a nice way to visualize this. We graph the values 0.1, $f(0.1)$, $f(f(0.1))$, and so forth. The horizontal axis shows the iteration number, n. At each step, we plot a point showing the value of $f^n(0.1)$. (The "zeroth" iteration is the start value, 0.1.) We join those points with line segments so the pattern is clearer. Here's the result:

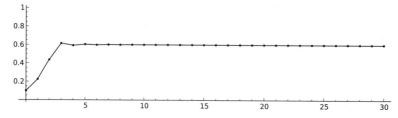

What we see is that the iterations of f converge to the value 0.6.

What's so special about 0.6? Notice that

$$f(0.6) = 2.5 \times 0.6 \times (1 - 0.6) = 2.5 \times 0.6 \times 0.4 = 0.6.$$

The value 0.6 is called a *fixed point* of f because running that number through the function f doesn't change it: $f(0.6) = 0.6$.

LET'S REPEAT THIS EXPERIMENT with another logistic map; this time we change the multiplier to 2.8; now the function is $f(x) = 2.8x(1 - x)$. As in the previous case, we start the iterations at $x = 0.1$. Here are the first ten:

n	$f^n(0.1)$
1	0.252000000000000
2	0.527788800000000
3	0.697837791264768
4	0.590408583372939
5	0.677113606546995
6	0.612166157052565
7	0.664772508993766
8	0.623980056783718
9	0.656961047455737
10	0.631017042828474

It looks like the iterations are dancing around 0.64. Let's run further iterations and graph the result:

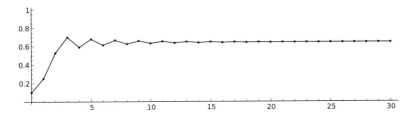

At the ten iteration mark the values are still moving up and down a bit, but by the time we reach 30 iterations they've settled down. To what value? It's a number between 0 and 1 with the property that $f(x) = x$. This is an equation we can solve like this:

$$f(x) = x$$
$$2.8x(1-x) = x \quad \text{divide both sides by } x$$
$$2.8(1-x) = 1$$
$$2.8 - 2.8x = 1$$
$$1.8 = 2.8x$$
$$x = \frac{1.8}{2.8} = 0.642857.$$

The iterations of $f(x) = 2.8x(1-x)$ converge to 0.642857.

THE ITERATIONS OF THE LOGISTIC MAP $f(x) = mx(1-x)$ can be considered as a simple evolving system. The number x represents the state of the system and the function f dictates how the system evolves from one "time" step to the next. In the two cases we've considered (with $m = 2.5$ and $m = 2.8$) the long-term behavior of the system is to settle down to an "equilibrium" at a fixed point of the function.

Mathematicians call these *dynamical systems*: systems with a starting state and a transition rule that describes how the state changes over time.

We continue our exploration of the iterations of the logistic map with the case $m = 3.2$. As in the prior cases, we start with $x = 0.1$. Here are the values of the iterates:

n	$f^n(0.1)$
1	0.288
2	0.6561792
3	0.721945783959552
4	0.642368220744256
5	0.735140127110768
6	0.623069185991463
7	0.751532721470076
8	0.597540128095543
9	0.769554954915536
10	0.567488404097546

What's going on? The iterates don't appear to be settling down. Indeed, if we look at the even steps, the values are getting smaller (roughly 0.66, 0.64, 0.62, 0.6, 0.57) and at the odd steps the values are getting larger (roughly 0.72, 0.74, 0.75, 0.77). Rather than coming together, the values are drifting apart!

Let's plot the first 30 iterations to attempt to visualize the behavior of this system:

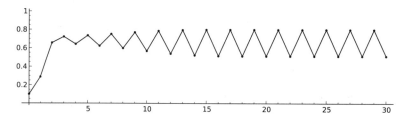

Behold! The system does not "settle down" to a single value, but rather it oscillates between two numbers. Let's run the computations out to the 50th iteration. Here are the last few lines of the chart:

n	$f^n(0.1)$
47	0.799455490467370
48	0.513044509532630
49	0.799455490467370
50	0.513044509532630

The long-term behavior of this system is an oscillation between two values, $s = 0.799455\ldots$ and $t = 0.5130445\ldots$. These numbers have the property that $f(s) = t$ and $f(t) = s$. The oscillation pattern is illustrated by the figure in the margin.

WHAT OTHER BEHAVIORS MIGHT WE OBSERVE when iterating a logistic map? The next stop on our tour is $m = 3.52$. Let's look directly at a graph of the iterates $f(x), f^2(x), f^3(x), \ldots$.

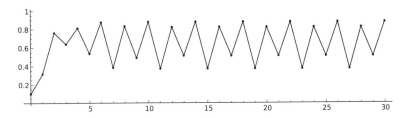

And here is a chart of the values:

n	$f^n(0.1)$
1	0.3168
2	0.7618609152
3	0.638629591038977
4	0.812352064439049
5	0.536575381199139
6	0.875291090045285
⋮	⋮
95	0.373084390547640
96	0.823301346832223
97	0.512076361760377
98	0.879486648432946
99	0.373084390487175
100	0.823301346778198

The long-term behavior of this system is interesting but quite stable. The system steps through a cycle of four different values, repeating them *ad infinitum*, as illustrated in the figure in the margin.

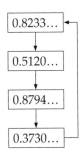

From order to chaos

WE HAVE EXAMINED THE LONG-TERM BEHAVIOR of the iterates of some logistic functions $f(x) = mx(1-x)$. In all cases, the iterates have settled down to a stable pattern. In some cases ($m = 2.5$ and $m = 2.8$) the system parked itself at a single value: a fixed point of the function f. In the other cases

238 THE MATHEMATICS LOVER'S COMPANION

($m = 3.2$ and $m = 3.52$) the system fell into a stable, predictable rhythm.

Life is good. We know the starting value: $x = 0.1$. And we know the rule that takes us from one value to the next: $f(x) = mx(1 - x)$. Of course we can work out from now until eternity the behavior of the system. Right?

Time for one final example: $m = 3.9$. Let's get the first handful of iterations from the computer:

n	$f^n(0.1)$
1	0.351
2	0.8884161
3	0.386618439717081
4	0.924864024972462
5	0.271013185108377
6	0.770503650562580
7	0.689628322626039
8	0.834760287106335
9	0.537948645688288
10	0.969383611132657

It's not clear what's happening. Let's try plotting the first 30 iterates.

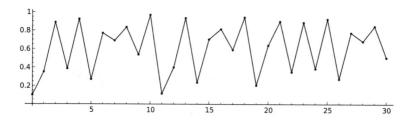

Hmmm.... There's no apparent pattern. No worries. Let's try again, this time running out to 100 iterations.

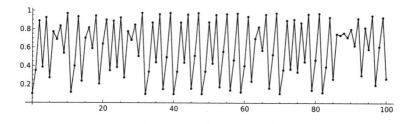

It looks random. Which, of course, it most certainly is not! Each value is precisely calculated from the previous by the function $f(x) = 3.9x(1-x)$. The iterates of this logistic map never "settle down" into a nice pattern. The irregularity continues forever.

FINE. THE ITERATES DON'T BECOME REGULAR but the system is 100% predictable.

- We know the initial state, $x = 0.1$.

- We know the transition rule to get from one state to the next, $x \mapsto 3.9x(1-x)$.

That means we can calculate exactly the state of the system after, say, one thousand iterations. Right?

Wrong.

We are stymied by the confluence of two problems: *round-off errors* and *sensitive dependence on initial conditions*. We discuss each in turn.

When we perform calculations on a calculator or a computer, the result we see is often an approximation. For example, when we divide $1 \div 3$, our device reports the answer as the decimal 0.3333333. The correct answer has infinitely many 3s, but the calculator only retains a handful of digits. After performing a few iterations of the function $f(x) = 3.9x(1-x)$, digits are lost about a dozen places to the right of the decimal point. That means that after just a few iterations, the computer reports an answer that is not exact; it's just an approximation. Normally we are not too concerned about such errors. If we are trying to calculate how much paint we'll need to cover a certain amount of wall space, the fact that the calculator might report an answer that's off by one part in a trillion is meaningless. Why should round-off errors matter in this case?

That brings us to the next problem: *sensitive dependence on initial conditions*. We calculate iterations of the function at two different, but nearly equal, starting values: 0.1 and 0.10001. Intuitively, we expect that tiny difference in starting value to be insignificant. Is it? Here's what happens.

n	$f^n(0.1)$	$f^n(0.10001)$
Start	0.1	0.10001
1	0.351	0.35103119961
2	0.8884161	0.88845235639
3	0.386618439717	0.386508590577
4	0.924864024972	0.92476682995
5	0.271013185108	0.271335246678
6	0.770503650563	0.771078479294
7	0.689628322626	0.688414186448
8	0.834760287106	0.836550367946
9	0.537948645688	0.533262014357
10	0.969383611133	0.970685189764
11	0.11574819984	0.110976263335
12	0.399167160889	0.384776076015
13	0.935347680371	0.923221444631
14	0.235842349062	0.276446074336
15	0.702860868258	0.780092205049
16	0.814505125705	0.669038591017
17	0.589237451031	0.863561223513
18	0.943943041601	0.459510623354
19	0.206366845676	0.968606380477
20	0.638740325658	0.118591434686

Notice that for the first dozen iterations or so, the system that started at 0.1 and the system that started at 0.10001 are in virtual lockstep. After that, however, their subsequent trajectories are rather different. This is nicely illustrated by plotting both evolutions on the same graph. The solid line is for the system starting at 0.1 and the dashed line is for the initial state 0.10001.

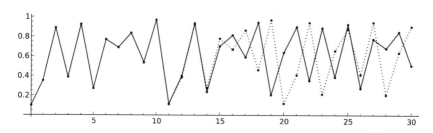

What is $f^{1000}(0.1)$? That is, if we perform one thousand iterations of $f(x) = 3.9x(1-x)$ starting at $x = 0.1$, what is the result?

We can do this calculation on a computer, but the results are nonsense. To illustrate this fact, we did this calculation three times using a different level of precision (number of digits the computer uses for numbers) and got the following results:

Level of precision	Value of $f^{1000}(0.1)$
Standard	0.967077
Double	0.395498
Quadruple	0.425983

Most likely *none* of these is the correct value for $f^{1000}(0.1)$.

One last-ditch effort. Computers can work with *arbitrary* precision. That is, we can request that no round off takes place from one iteration to the next. But here's what happens when we do that:

Iteration (n)	$f^n(0.1)$
Start	0.1
1	0.351
2	0.8884161
3	0.386618439717081
4	0.924864024972462138613439673812
5	0.2710131851083763662000140866401248824984463548094469784701850 01

The exact value of $f^6(0.1)$ is 127 digits long and $f^7(0.1)$ is 255 digits. The pattern is that the number of digits we need to express $f^n(0.1)$ roughly doubles in each iteration. No computer is large enough to calculate $f^{1000}(0.1)$.

WHERE DOES THIS LEAD US? Although we know the starting point of the system and the rule that takes us from one state to the next, we can't figure out the state of the system after 1000 steps.

One can prove that $f^{1000}(0.1)$ lies between 0 and 1, and so we might ask: What is the probability that $f^{1000}(0.1)$ is, say, greater than $\frac{1}{2}$?

The answer to that question is either 0 or 1 because there's nothing random here. Either $f^{1000}(0.1) > \frac{1}{2}$ or else $f^{1000}(0.1) \le \frac{1}{2}$. There's no maybe, no randomness.

This simple system is *chaotic*. It is absolutely deterministic and utterly unpredictable.

There are many mathematical systems that exhibit chaotic behavior, and many of these arise from models of the real world, such as the weather.

The Collatz $3x + 1$ problem

SO FAR, THIS CHAPTER HAS ONLY CONSIDERED iterations of logistic maps. We end with a rather different type of function and a thorny, unsolved problem about its iterates.

The logistic map is a function given by a simple algebraic formula. However, it is possible to define functions by other means. The function F we consider now is defined only for positive integers and is defined by the following rule:

$$F(x) = \begin{cases} 3x+1 & \text{if } x \text{ is odd and} \\ x/2 & \text{if } x \text{ is even.} \end{cases}$$

This function involves two simple algebraic formulas, but the formula we invoke depends on whether x is even or odd.

Some examples:

- $F(9) = 28$. The number 9 is odd, so we apply the formula $3x + 1$ to calculate $3 \times 9 + 1 = 28$.

- $F(10) = 5$. The number 10 is even, so we apply the formula $x/2$ and we get $10/2 = 5$.

When x is odd, $F(x)$ is a positive integer. Also, when x is even, $F(x)$, which equals $x/2$, is again a positive integer.

Succinctly: If x is a positive integer, so is $F(x)$.

This means that we can iterate F because the output of F is a valid input for F. Let's illustrate what happens if we iterate F starting with $x = 12$.

- $F(12) = 6$ because 10 is even.
- $F^2(12) = F(6) = 3$ because 6 is even.
- $F^3(12) = F(3) = 10$ because 3 is odd, and so we use the $3x+1$ formula.
- $F^4(12) = F(10) = 5$.

Here's a handy way to illustrate the iterations. We write $12 \mapsto 6$ to show that applying F to 12 results in the value 6. We can now track the iterations of F like this:

$$12 \mapsto 6 \mapsto 3 \mapsto 10 \mapsto 5 \mapsto 16 \mapsto 8 \mapsto \underbrace{4 \mapsto 2 \mapsto 1} \mapsto \underbrace{4 \mapsto 2 \mapsto 1}.$$

Notice that the pattern $4 \mapsto 2 \mapsto 1$ is repeated. What happens if we keep going? Since $F(1) = 4$, $F(4) = 2$, and $F(2) = 1$, the next three terms are also 4, 2, and 1.

In other words, if the sequence of iterates ever gets to 1, then forever after the pattern will be $4, 2, 1, 4, 2, 1, 4, 2, 1 \ldots$.

Let's try a different start value, say 9. Here's what we get: $9 \mapsto 28 \mapsto 14 \mapsto 7 \mapsto 22 \mapsto 11 \mapsto 34 \mapsto 17 \mapsto 52 \mapsto 26 \mapsto 13 \mapsto 40 \mapsto 20 \mapsto 10 \mapsto 5 \mapsto 16 \mapsto 8 \mapsto 4 \mapsto 2 \mapsto 1$.

Here's an impressive set of iterations: $27 \mapsto 82 \mapsto 41 \mapsto 124 \mapsto 62 \mapsto 31 \mapsto 94 \mapsto 47 \mapsto 142 \mapsto 71 \mapsto 214 \mapsto 107 \mapsto 322 \mapsto 161 \mapsto 484 \mapsto 242 \mapsto 121 \mapsto 364 \mapsto 182 \mapsto 91 \mapsto 274 \mapsto 137 \mapsto 412 \mapsto 206 \mapsto 103 \mapsto 310 \mapsto 155 \mapsto 466 \mapsto 233 \mapsto 700 \mapsto 350 \mapsto 175 \mapsto 526 \mapsto 263 \mapsto 790 \mapsto 395 \mapsto 1186 \mapsto 593 \mapsto 1780 \mapsto 890 \mapsto 445 \mapsto 1336 \mapsto 668 \mapsto 334 \mapsto 167 \mapsto 502 \mapsto 251 \mapsto 754 \mapsto 377 \mapsto 1132 \mapsto 566 \mapsto 283 \mapsto 850 \mapsto 425 \mapsto 1276 \mapsto 638 \mapsto 319 \mapsto 958 \mapsto 479 \mapsto 1438 \mapsto 719 \mapsto 2158 \mapsto 1079 \mapsto 3238 \mapsto 1619 \mapsto 4858 \mapsto 2429 \mapsto 7288 \mapsto 3644 \mapsto 1822 \mapsto 911 \mapsto$

2734 ↦ 1367 ↦ 4102 ↦ 2051 ↦ 6154 ↦ 3077 ↦
9232 ↦ 4616 ↦ 2308 ↦ 1154 ↦ 577 ↦ 1732 ↦
866 ↦ 433 ↦ 1300 ↦ 650 ↦ 325 ↦ 976 ↦ 488 ↦
244 ↦ 122 ↦ 61 ↦ 184 ↦ 92 ↦ 46 ↦ 23 ↦ 70 ↦
35 ↦ 106 ↦ 53 ↦ 160 ↦ 80 ↦ 40 ↦ 20 ↦ 10 ↦
5 ↦ 16 ↦ 8 ↦ 4 ↦ 2 ↦ 1.

The sequence eventually reaches 1, but it takes more than one hundred iterations.

The Collatz $3x+1$ conjecture states that no matter what the initial positive integer x might be, the sequence of iterates $x \mapsto F(x) \mapsto F^2(x) \mapsto F^3(x) \mapsto \cdots$ eventually reaches 1, and from there the pattern is 4, 2, 1 forever more.

This problem was created by Lothar Collatz in 1937.

This problem has defied solution despite extensive attention from professional and amateur mathematicians alike.

Question from page 232: We have $f(x) = x^2 + 1$ and $g(x) = 1 + x + x^2$. So

$$g(f(2)) = g(2^2 + 1) = g(5) = 1 + 5 + 5^2 = 31.$$

22
Social Choice and Arrow's Theorem

DEMOCRACY IS A PROCESS in which decisions are made based on the opinions of the members of a society. This is accomplished by giving individuals an opportunity to express their preferences (by voting) and then combining those individual preferences to produce a decision.

Two-party elections

THE MOST FAMILIAR INSTANCE OF THIS PROCESS is an election in which two candidates vie for the same office. Members of the community cast ballots for one of the two candidates, and whoever gets the most votes wins.

The key phrase is: *whoever gets the most votes wins*. This is the cornerstone on which democratic societies are built. But is it fair?

Imagine that the two candidates running for office are named A and B. The ballot we present is simple: The voters mark which of these two candidates they prefer. If there are n voters, the data we collect looks like this:

Voter	#1	#2	#3	......	#n
Preference	A	A	B	......	A

How do we transform this preference profile into

A more complicated ballot might allow voters to indicate how strongly they prefer one candidate over another.

We use the phrase *preference profile* to mean the entire collection of individual preferences.

a decision? The usual method is to count how many people prefer A and how many prefer B. The winner of the election is the candidate with the most votes. We call this method MAJORITY and it's the method of choice for democratic societies. But this is not the only possible method for converting a preference profile into a decision. Let's look at some alternatives.

The DICTATOR method says that the result of the election is the preference of a specific individual, say, Voter #1. If #1 prefers A, A wins, but if #1 prefers B, then B wins. No one else's opinion matters.

We call MAJORITY and DICTATOR *decision methods*. Both take, as input, a preference profile, and then return, as output, a decision. In the real world, both methods are used, but in general we think of DICTATOR as being unfair. Why?

To be considered fair, a decision method should satisfy certain properties. The offensive aspect of DICTATOR is that it does not treat all votes equally. More formally, a fair decision method should exhibit *voter neutrality*—this is just a fancy way of saying that it doesn't matter *who* holds which preferences, but only how many people hold each of the preferences. The MAJORITY decision method satisfies the *voter neutrality* property, but DICTATOR does not.

If we only use *voter neutral* decision methods, we can summarize the preference profile just by giving a tally of how many voters prefer each candidate. The profile can be neatly presented in a chart like this:

A	B
23	17

Here's another decision method we call ALPHABETICAL. In this method, whichever candidate's name comes first in alphabetical order is the winner. That means, for the election of A versus B, that A is the winner.

Clearly this decision method is unfair, but why?

It's important not to confuse decision methods (such as MAJORITY) with properties those methods might have (such as *voter neutrality*). To highlight the difference we use SMALL CAPITALS for decision methods and *italics* for properties of those methods.

It satisfies the *voter neutrality* property: it treats all voters exactly the same by completely ignoring everyone's preference! The problem is that it does not treat the candidates equally. We say that a decision method satisfies *candidate neutrality* when the candidates are treated equivalently; changing the names of the candidates doesn't affect the outcome.

Notice that the DICTATOR method satisfies the *candidate neutrality* property.

Our sense of fairness leads us to require that a decision method should exhibit both *voter neutrality* and *candidate neutrality*. Is this enough?

Here's another decision method we call ODD: Whichever candidate is preferred by an odd number of voters is the winner. So if A is preferred by 20 voters and B is preferred by 13 voters, then B is the winner. This method is both *voter neutral* and *candidate neutral*.

Or consider the MINORITY method: Whichever candidate is preferred by the fewest number of voters wins. So if A is preferred by 12 voters and B is preferred by 30, then A wins. This method is also both *voter* and *candidate neutral*.

The two properties *voter neutrality* and *candidate neutrality* rule out some unfair decision methods (such as DICTATOR and ALPHABETICAL) but some silly methods (such as ODD) meet both these requirements. So let's introduce another property that can help us distinguish sensible methods (such as MAJORITY) from silly methods.

Here's a problem with the ODD decision method. Suppose the preference profile looks like this:

A	B
15	10

Profile I

By the ODD decision method, A is the winner.

Now let's suppose that one of the voters has a change of heart and switches preference from B (the loser) to A (the winner). Only this one voter switches opinion; all the others keep their preferences as before. The result is this new preference profile:

A	B
16	9

Profile II

By the ODD decision method, the winner is now B.

Now that's odd: If a voter changes his/her opinion from favoring the loser to favoring the winner that shouldn't hurt the winner, but that's exactly the situation here. The ODD decision method violates the *monotonicity* property.

There's another problem with the ODD decision method. What happens if there is an even number of voters? We could have one of the following two situations:

A	B
18	12

A	B
17	13

Applying the ODD decision method to the first profile yields no winner and applying it to the second profile results in two winners. Either way, it's a tie.

It would be desirable if a decision method avoided ties so the collective opinion of the voters is guaranteed to produce a definitive decision. Some decision methods (such as DICTATOR) won't produce a tie. But a decision method that is both *voter* and *candidate neutral* must declare a tie if the electorate is equally divided.

Thus, if we require our decision method to be *voter* and *candidate neutral* we have to accept the fact that if exactly half the voters prefer one candidate and exactly half prefer the other, then a tie must be declared. And this is the case with the MAJORITY decision method.

However, MAJORITY fails to pick a unique winner only in this situation. We say that MAJORITY is *almost decisive* meaning that a unique winner is always

Here's a formal statement of *monotonicity*. To say that a decision method is *monotone* means that if the method declares a winner X from a given preference profile, and if one voter changes his/her preference from the loser to X to create a new profile (with just this one change), then the decision method must pick X to be the winner in the new profile. In other words, switching one voter's preference from the loser to the winner doesn't hurt the winner.

By a technicality, DICTATOR is *almost decisive* because there can never be a tie.

	Voter Neutral	Candidate Neutral	Monotone	Almost Decisive
MAJORITY	Yes	Yes	Yes	Yes
MINORITY	Yes	Yes	No	Yes
DICTATOR	No	Yes	Yes	Yes
ALPHABETICAL	Yes	No	Yes	Yes
ODD	Yes	Yes	No	No

Voting methods and their properties.

chosen except, perhaps, in the case that the voters are exactly split in their preference.

The MINORITY voting system is also *almost decisive* (but is not *monotone*).

We have identified four properties of decision methods that capture our sense of fair play; a fair decision method should be *voter neutral, candidate neutral, monotone,* and *almost decisive*. And, happily, MAJORITY satisfies all four of these properties. The table above summarizes the properties held by the voting methods we have considered thus far.

But perhaps there's a viable alternative. Is there another decision method that has all four of these attributes?

The answer is no. In 1952, Kenneth May proved that MAJORITY is the one and only decision method that is *voter neutral, candidate neutral, monotone,* and *almost decisive*.

Elections with more than two candidates

OUR INTUITION THAT MAJORITY RULE is a fair way to combine individual opinions to make a decision is backed up by mathematical rigor. For a two-party election, May's theorem shows that it's the only reasonable choice.

The situation changes considerably when there are more than two candidates. Still, we may hope that a method akin to MAJORITY will emerge as the only

In this section we only consider elections that should result in a single winner. It's not unusual for there to be elections in which two or more winners are to be selected from a larger slate of candidates. A common instance of this is a school board election.

viable system.

Let's begin with the method by which voters express their preferences. When there are three (or more) candidates, we assume that each voter rank orders the candidates from most desirable to least. A preference profile in this situation looks like this:

Voter	#1	#2	#3	......	#n
1st choice	A	A	B		D
2nd choice	B	C	A		A
3rd choice	D	B	D		B
4th choice	C	D	C		C

> We're going to keep things as simple as possible. We can imagine that a voter prefers, as first choice, A, is indifferent between B and C, and really despises D. Nevertheless, we would expect this voter to rank order the candidates and don't provide a mechanism for specifying the intensity of the opinions.
>
> Mathematicians do consider more complicated ways for voters to specify preference, but we'll stick with the simplistic rank-order method.

As before, we want to find a decision method that takes, as input, a preference profile and returns a winner as its output.

For example, the DICTATOR method specifies that the first choice candidate of, say, voter #1 is the winner of the election. For the profile just presented, that would be candidate A; the opinions of all the other voters are ignored.

The DICTATOR method fails to satisfy *voter neutrality* (but does satisfy *candidate neutrality*). Since (presumably) we only want to consider *voter neutral* methods, a preference profile can be reported by counting how many people have each of the possible preference orderings. For example, if there are three candidates the preference profile can be summarized like this:

A	A	B	B	C	C
B	C	A	C	A	B
C	B	C	A	B	A
8	12	3	11	9	0

> If there are k candidates running for office, a preference profile would have $k!$ columns; see Chapter 10.

In this profile 20 people have A as their first choice, 14 like B best, and 9 see C as the best alternative. What method should we use to choose the winner?

MAJORITY IS THE METHOD OF CHOICE when there are two candidates. When there are three or more candidates there can be a majority result if more than half the voters share the same first choice. This doesn't necessarily happen, and so extending MAJORITY to this more general setting can be problematic. Also, MAJORITY does not take into account voters' second or lower choices. Let's see why this matters. Consider the following preference profile:

A	A	C	D
B	B	B	B
C	D	D	C
D	C	A	A
12	12	10	10

Notice that more than half of the voters prefer A as their first choice. Is it clear that choosing A is the "best" choice? What might we mean by "best"? These are not mathematical questions, but our values can guide how to develop notions of fairness. To illustrate this, let's suppose the candidates are restaurants and the voters are office workers selecting a venue for a group dinner. Here's a bit more information about the restaurants:

Restaurant	Description
A	Steak house
B	Extensive buffet
C	Indian food
D	Greek food

The picture comes into focus. Most of the office workers (24) prefer a steak dinner but a good number of them (20) strongly dislike this choice. The latter group prefers Indian or Greek food as their first choice.

What we see is that everyone in the office picks the buffet restaurant as his/her second choice. This seems like a good compromise, and a wise boss would likely choose this location for the company dinner. Is it possible to create a decision method that would reach the same conclusion for elections?

> ### Preference Profiles versus Ballots
>
> We haven't discussed *how* voters express their preferences; we just assume we know how each voter ranks the candidates. The preference profile is a tally of all rankings from all the voters.
>
> In a typical election, the ballot only lets voters select a single candidate; it does not present voters with a way to rank order those running for office. That makes sense if the decision method is PLURALITY as that method only considers voters' top choices.
>
> Occasionally one sees ballots that allow voters to select more than one candidate. If the decision method is VOTE-FOR-TWO, then the voters need to report their top two choices, and they need not distinguish which candidate they prefer.
>
> In this chapter we assume that each voter has a preference order and that the ballot used captures enough information about voter preferences to suit the decision method being used. So for the DICTATOR method, we need not bother collecting ballots from anybody (except the dictator!) but for the BORDA method (described on the facing page) to be applied we would need all voters to complete a ballot with their full ranking.
>
> In other words, once we decide what decision method we are going to use, we would then design a ballot to capture enough information from voters to apply that method.

THERE'S A PLETHORA of decision procedures for multiple-candidate elections. While MAJORITY is ideal for two-party elections, it's not a viable choice more generally both because it is often the case that no single candidate will be the first choice of more than 50% of the voters and, as our restaurant example illustrates, because it is not clear that this choice is the "right" decision.

So let's present a few decision methods and then try to figure out which might be the best. We'll use this voter profile to illustrate how the three methods work.

A	C	B	A	B
C	B	A	B	C
B	A	C	C	A
4	2	1	2	4

Preference Profile for Three Methods

- PLURALITY. This is the method that might be most familiar. We choose whichever candidate is the first choice of the most voters. We don't require that more than half of the voters agree in this regard.

  For the preference profile on the current page we note that A is the first choice of the most voters (six), then B is next (five), and C is last (two). So applying the PLURALITY method to this profile we would declare A to be the winner.

- VOTE-FOR-TWO. One problem with the PLURALITY method is that it ignores voters' preferences beyond which they like best.

  The VOTE-FOR-TWO method counts how many times each candidate is the first or second choice of the voters. For the preference profile on this page we get the following results:

  - A gets $6 + 1 = 7$ votes (six first place and one second place).
  - B gets $5 + 4 = 9$ votes (five first place and four second place).
  - C gets $2 + 8 = 10$ votes (two first place and eight second place).

  Hence C wins when we use the VOTE-FOR-TWO method.

- BORDA. The PLURALITY decision method gives no credence to voters' second choices while, perhaps, the VOTE-FOR-TWO gives too much importance to second choices by giving them the same weight as first choices. The BORDA method is a compromise between these two.

In this method, each candidate gets 2 points for being a voter's top choice, 1 point for being a voter's second choice, and no points for being the voter's last choice. We then add up the points, and the candidate with the most points wins.

Let's analyze how this works for the preference profile on the previous page.

- Candidate A is the top choice for 6 people and the second choice for one. Therefore we assign

  points for A → $6 \times 2 + 1 \times 1 = 13$.

- Candidate B is the top choice for 5 people and the second choice for 4. Therefore we assign

  points for B → $5 \times 2 + 4 \times 1 = 14$.

- Candidate C is the top choice for 2 people and the second choice for 8. Therefore we assign

  points for C → $2 \times 2 + 8 \times 1 = 12$.

We see that the BORDA method declares B to be the winner.

> This method is named for Jean-Charles de Borda, an eighteenth-century French mathematician.
>
> In de Borda's method, if there are four candidates, then the top choice candidate gets 3 points, second choice candidates gets 2 points, third choice gets 1 point, and last place gets 0 points. Point values for five candidates are 4, 3, 2, 1, 0, and so on. Notice that the BORDA method in the case of two candidates is the same as MAJORITY.

Here's a summary of the methods and the decisions they reach, all for the same preference profile:

Method	Winner
PLURALITY	A
VOTE-FOR-TWO	C
BORDA	B

This is unfortunate. It's hard to fault any one of these methods as being ridiculous (like the ODD or MINORITY methods). All three of these methods satisfy the fairness criteria of *voter neutrality, candidate neutrality*, and *monotonicity*, so we can't use one of those properties to invalidate any of these methods. Perhaps we can find another fairness criteria to guide our choice for the "best" decision method.

The independence of irrelevant alternatives

THE FINAL FAIRNESS CRITERION we present is called the *independence of irrelevant alternatives* condition. It's more technical than the previous criteria we presented, so here's a rough analogy.

Imagine your friend is ordering dessert after dinner at a restaurant. The choices are cake, pie, or ice cream. Your friend orders ice cream, but just after placing her order, the waiter tells you, "Oh, it seems we're out of pie." At this point, your friend says, "In that case, I'll have cake!"

Does that make any sense? If ice cream is the your friend's top choice (from among cake, pie, and ice cream) then the fact that the restaurant is out of pie should be irrelevant. We're assuming here that your friend's change of mind was *because* pie is unavailable (and not merely a sudden change of preference). We might be tempted to question our friend's sanity!

We should expect sanity for decision methods. Suppose a decision method declares X to be the winning candidate for some preference profile. Now imagine that some other candidate, say Y, drops out of the race (and no voter changes his/her preference). In this case, X should still be the winner. A decision method that works this way is said to satisfy the *independence of irrelevant alternatives* condition.

Let's think about the PLURALITY method. For the preference profile on page 253, the PLURALITY method declares A to be the winner. Now suppose C drops out of the contest. The preference profile changes like this:

A	C	B	A	B
C	B	A	B	C
B	A	C	C	A
4	2	1	2	4

$\Longrightarrow$

A	B	B	A	B
B	A	A	B	A
4	2	1	2	4

$\Longrightarrow$

A	B
B	A
6	7

Now B is the winner! The PLURALITY method does not satisfy the *independence of irrelevant alternatives* condition.

Perhaps VOTE-FOR-TWO does better. For the preference profile we have been examining, VOTE-FOR-TWO chooses C as the winner. What happens if A drops out of the race? There are only two candidates left! This means we have a tie. As a puzzle, try to create a preference profile in which there are four candidates (A, B, C, and D), VOTE-FOR-TWO declares A to be the winner, but when D drops out, then B is the winner. A solution appears on page 258.

Finally, let's consider the BORDA method. It declares B to be the winner for this preference profile, but if, as before, C drops out, then A wins by the BORDA method.

None of the three methods we presented satisfy the *independence of irrelevant alternatives* condition.

NOT TO WORRY! There are *many* more decision methods that people have proposed. Surely one of them satisfies the *independence of irrelevant alternatives* condition. For example, DICTATOR satisfies this condition. (If A is the first choice of voter #1, then A wins whether or not another candidate drops out.) Of course, DICTATOR is not an acceptable choice for a decision method as it violates *voter neutrality*.

Still, it makes sense to ask: What fair decision method is there that satisfies *independence of irrelevant alternatives*? The answer was found definitively in 1950 by Kenneth Arrow. Unfortunately, the definitive answer is: there are none.

Arrow's Impossibility Theorem is a bit technical, but it implies that, when there are three or more candidates, no decision method can satisfy these basic fairness criteria.

What are we to do? Given that there are no "fair" decision methods, which one should we pick? Or is the *independence of irrelevant alternatives* requirement simply irrelevant? That is to say, is there any harm in choosing a decision method that does not satisfy this criterion?

The problem with adopting a decision method that does not satisfy the *independence of irrelevant alternatives* requirement is that it can encourage voters to report a ranking that differs from their preferences. This happens in the face of a "spoiler." Suppose you like candidates A and B, but really dislike C. You have a preference for A, but you suspect (from news reports) that A does not have much of a chance of winning. However, both B and C are strong candidates. Whom do you report as your top choice? In PLURALITY voting (and other methods) it may be unwise to vote for A despite A being your top choice. In essence, A takes away your vote from B.

If A stays in the race and people that think like you cast their top vote for A, that may deplete votes from B and hand a victory to C. But were A to drop out of the race, your vote would go to B, and perhaps give B the edge needed to win.

If the decision method used to decide this election satisfied the *independence of irrelevant alternatives* you would not face this dilemma. You could accurately record your preferences knowing that ranking A above B would not "waste" your vote.

Showing that VOTE-FOR-TWO does not satisfy the *independence of irrelevant alternatives* property.

Consider this preference profile:

A	A	B
B	D	C
C	B	D
D	C	A
5	5	4

The VOTE-FOR-TWO decision method gives points as follows:

Candidate	A	B	C	D
First place	10	4	0	0
Second place	0	5	4	5
Total votes	10	9	4	5

We see that A wins this election.

Now suppose D drops out of the competition. The preference profile now looks like this:

A	A	B
B	B	C
C	C	A
5	5	4

Now VOTE-FOR-TWO gives 10 votes to A, 14 votes to B, and 4 votes to C, so B is the winner.

In essence, D is a spoiler, robbing votes from B (in the second column).

Thus, initially A is the winner but when the "irrelevant alternative" D drops out, the winner switches to B. This shows that VOTE-FOR-TWO does not satisfy the *independence of irrelevant alternatives* property.

23
Newcomb's Paradox

HUMAN BEHAVIOR IS PREDICTABLE. Indeed, much of social science, from economics to cultural anthropology, relies on the fact that we can see patterns in human activity and foresee what people will do (albeit with varying degrees of certainty).

In this chapter we present Newcomb's Paradox, which takes prediction of human behavior to a mind-bending conclusion.

This conundrum was conceived by William Newcomb and appeared first in a scholarly article by Robert Nozick. It gained a greater audience thanks to Martin Gardner's regular column in *Scientific American*.

Newcomb's game

NEWCOMB'S PARADOX TAKES THE FORM OF A GAME involving two players: the Predictor and the Chooser. In this game, the Chooser is presented with two options that involve winning money. However, prior to making a decision, the Predictor guesses what the Chooser is going to choose. To make this more personal, imagine that you, the reader, are playing the role of Chooser.

Imagine that before you sit two boxes named Box #1 and Box #2. The boxes are opaque so you can't tell what they contain. However, we guarantee that Box #1 contains $1,000. The situation is murkier for Box #2. It might contain $1,000,000 or it might be empty. The situation looks like this:

Box #1	Box #2
$1,000 guaranteed	$1,000,000 or $0

In a moment we will explain when it is empty and when it holds the million dollars. First, we present you with your choices. When it is your turn, you may do exactly one of the following:

- Take the contents of both boxes.

- Take the contents of Box #2 only.

And when we say "take the contents" of course we mean you get to keep the money.

However, the Chooser does not make the first move. Before you render your decision, the Predictor tries to figure out what you're going to do. We imagine the Predictor is an enormously talented psychologist who has silently observed you for years, can remotely measure your heart rate and perspiration, and, based on all these inputs, can get a really good idea of what you're going to do. The Predictor, we should hasten to add, is not infallible; that would be unrealistic. For now, let's say the Predictor is simply very good, by which we mean, is right 95% of the time.

Now let's explain how the boxes are loaded with cash. As promised, Box #1 contains a thousand dollars. The contents of Box #2 depends on the Predictor's forecast of your actions. If the Predictor thinks you will grab both boxes (which is your right), then Box #2 is left empty. But if the Predictor thinks you will take Box #2 only, then that box is stuffed with one million dollars. In table form, the link between predictions and the contents of the boxes is as follows:

Perhaps you think the 95% accuracy figure is also too accurate. That's fine; later we will lower the estimate to 51% and this will not affect the result. So please, for now, accept the 95% accuracy figure.

There is one more technical rule that we'll explain in a moment.

If the Predictor thinks ...	Then Box #2 contains ...
you will take Box #2 only	$1,000,000
you will take both boxes	$1,000

Let's recap.

1. The Predictor goes first.

   - If the Predictor thinks the Chooser (you) will take just Box #2, then the million dollars is placed in that box.
   - However, if the Predictor thinks the Chooser will take both boxes, then Box #2 is left empty.

2. Then it's the Chooser's turn. You may do one of the following two actions:

   - Take Box #2 only.
   - Take both boxes.

In either case, you get the money in the boxes you pick!

HERE'S THE QUESTION WE WISH TO ANALYZE:

> *What should the Chooser do in order to get the most money?*

We need to distinguish this from a question that, superficially, seems to be the same: *What would you do in this situation?* We can imagine people responding in one of these ways:

- *I would only take the second box; I don't want to appear greedy.*

- *I would take both boxes. That way I know that at least I'd get some money.*

These are reasonable thoughts, but neither is an answer to the question we asked. They aren't (and don't try to be) strategies to get the most money we possibly can. However, that's the problem we're trying to solve. What should you do to get the most money possible?

Here is one more answer and, because this is also a reasonable idea, we're going to have a last-minute minor change to the rules to make this unviable.

- *I'd just flip a coin to decide.*

This throws a bit of a monkey wrench into the situation because the Predictor can't be expected to know the outcome of a coin flip. We therefore mildly modify the game simply to say that it must be your own decision what to do. You can't ask someone else to choose for you, you can't flip a coin or draw a random card from a deck, or do anything other than make up your own mind.

It boils down to this: What choice should you make to try to get the most money out of this game?

The good news is that there's a correct answer to this problem. The bad news is that there's more than one correct answer. Let's see why.

Don't leave money on the table!

WITHOUT ANY DOUBT taking both boxes is a thousand dollars better than just taking Box #2. Remember: The Predictor goes first. While he or she is nearly always correct, you don't know what the Predictor has done. The million dollars either is or is not in Box #2. If you only take Box #2, you'll get whatever sum is in that box. But if you take *both* boxes you'll get whatever is in Box #2 *plus an additional thousand dollars*. Without doubt, that's more money.

Here's a way to visualize your options depending on whether or not the million dollars is in Box #2.

If you don't want the extra $1000 that's fine. Donate it to charity! Just don't leave it sitting on the table.

State of Box #2	Choose Box #2 only	Choose both boxes
Empty	$0	$1000
Full	$1,000,000	$1,001,000

It's now very clear. Regardless of what the Predictor thinks or does, you are *always* better off taking both boxes.

Greed doesn't pay

YOU'RE WAITING ON LINE TO PLAY THIS GAME. A bunch of your friends have already had their turns. Some of them heeded the infallible logic presented above and chose both boxes. Others decided, even in the face of irrefutable reasoning, just to take Box #2. What's the result?

The logical, rational players have a nice extra $1000, but the ones who ignored reason are now millionaires! Well, there was one person the Predictor got wrong, but for the rest of the previous players, the predictions were spot on.

Now it's your turn.

If you agree with the reasoning in the previous section, you're going to take both boxes; after all, why are you leaving the extra $1000 behind? If you do that, it's quite likely the Predictor anticipated that (because that's the sort of person you are) and left Box #2 empty. Too bad for you, because you have a very nice consolation prize (a thousand dollars) but you're not a millionaire.

On the other hand, if you think: That Predictor is really quite good. I've decided just to take Box #2 and then have good reason to hope that the Predictor anticipated that. Indeed, if that's the decision you make, you are very likely to be suddenly a lot richer!

In other words, the players that chose the Box #2-only strategy are getting a lot more money than the "logical" players that took both boxes. The evidence is overwhelming: Taking Box #2 only is *much more likely* to result in a large payoff whereas taking both boxes is *much more likely* to result in a significantly smaller prize.

Taking Box #2 only is the best way to the largest possible reward!

WE CAN RECAST THIS ARGUMENT using the language of expected value. This is a simple way to calculate how much money, on average, the different strategies (take Box #2 only versus take both boxes) will pay. We'll return to the Predictor-Chooser game in a moment, but to keep things simple, we consider a Pick Three lottery game.

In a Pick Three lottery game, players purchase tickets for $1 each. The customer picks a three-digit number for each ticket. Later that evening, there's a random drawing that generates the winning three-digit number. Every ticket that matches that winning number can be cashed in for $500.

We ask: What's the expected payoff for a Pick Three lottery ticket?

A sensible, colloquial answer is: Nothing. The probability that we correctly guessed the winning number is one in a thousand, or 0.1%. Nearly always (but not never) the ticket is a loser.

But when mathematicians ask about the expected payoff, they mean this: What is the *average* payoff of a Pick Three ticket? Here's how we calculate this.

With probability 0.999 the payoff is $0 and with probability 0.001 the payoff is $500. Therefore, on average, a ticket pays

$$\$0 \times 0.999 + \$500 \times 0.001 = \$0.50.$$

The value of a Pick Three lottery ticket is, on average, just 50 cents.

Here's another way to think about the same result. Imagine that the lottery agency sells a million Pick Three tickets. Since each ticket costs $1, they collect a million dollars. How much do they pay out?

When the winning number is picked, we should anticipate that one ticket in a thousand is a winner. That means there will be around 1000 winning tickets, and the agency pays $500 each. The total payout is therefore $500,000. On a per ticket basis, that comes to 50 cents per ticket sold.

LET'S APPLY THIS ANALYSIS to Newcomb's game. As Chooser, we may either take just Box #2 or we may take both boxes. What's the expected payout in each situation?

- If we take Box #2, there's a 95% chance the Predictor correctly anticipated that (and put the million in the box) and a 5% chance the Predictor is wrong (and left the box empty). Therefore, the expected payout is

    $$\$1{,}000{,}000 \times 0.95 + \$0 \times 0.05 = \$950{,}000.$$

- If we take both boxes, again there's a 95% chance this was correctly predicted (and Box #2 is empty) and 5% chance of an error (and Box #2 contains a million dollars). In both cases, the Chooser also gets the $1,000 in Box #1. Here's the expected payout:

    $$\$1{,}000 \times 0.95 + \$1{,}001{,}000 \times 0.05 = \$51{,}000.$$

It's now very clear. The best way to get the most money is to take Box #2 only.

Conflict and resolution

WE'VE REACHED TWO INESCAPABLE CONCLUSIONS. One: It's better to take both boxes (why leave money on the table?). Two: It's better to just take Box #2 (you're much more likely to be a millionaire if you do). How is this possible? These can't both be right!

What we have is a contradiction. We've dealt with contradictions before. In Chapter 1 we created a number N that (a) is divisible by a prime and (b) is not divisible by a prime. Of course, that's impossible. What enabled us to deduce a contradiction is that we made an erroneous assumption. Indeed, we intentionally assumed that there are only finitely many primes, and that supposition led us to an impossible pair of conclusions. Since the conclusion is impossible, the supposition was in error. It's not true that there are finitely many primes because that leads to utter nonsense. Therefore there are infinitely many primes.

For Newcomb's paradox we did not explicitly announce any assumptions, but indeed there are two.

First, there is a Chooser. Is it possible for a Chooser to make this decision? Some may question whether or not human beings have free will. Indeed, it may be impossible (and we won't try here) to give a definitive answer to this age-old philosophical problem.

> Just for the record: You have free will. I'm not sure why you might have decided to think you don't!

Second, there is a Predictor. Is it possible for a Predictor to accurately forecast human behavior? It's very clear that human behavior in aggregate can be predicted to a high degree of accuracy. But in this game the Predictor is trying to ascertain the action of an individual, and that's a rather different matter. The 95% accuracy value seems absurdly high.

Now comes the interesting part: The contradiction doesn't vanish when we say that the Predictor is right 51% of the time! The "don't leave money on the table" argument remains valid. Here's the expected value argument:

- If you take only Box #2, your expected return is

  $\$1{,}000{,}000 \times 0.51 + \$0 \times 0.49 = \$510{,}000.$

- If you take both boxes, your expected return is

  $\$1{,}000 \times 0.51 + \$1{,}001{,}000 \times 0.49 = \$491{,}000.$

> Here's a puzzle for you: Find the accuracy level at which the contradiction goes away. That is, for what Predictor accuracy level do the two options (Box #2 only versus both boxes) give the same expected outcome? The solution is on page 268.

Even if the Predictor is a mere 51% accurate, you are better off taking only Box #2! The margin is thin, so the emotional pull of the argument is muted, but the result still indicates that choosing Box #2 only is superior to taking both boxes.

Here, again, are the two hidden assumptions:

- The Chooser has free will.

- The Predictor can forecast the Chooser's decision with 51% (or better) accuracy.

What this suggests is that free will and even modest levels of predictability are incompatible.

By changing the two payoff amounts (one thousand and one million dollars) we can manipulate the required level to be a value as close to 50% as we like).

Computer as the Chooser

LET'S IMAGINE THAT A COMPUTER PROGRAM plays the role of the Chooser in Newcomb's game, and we (human beings) play the role of Predictor. By the rules of the game, the Chooser is not permitted to flip a coin to decide; this means that the computer program playing Chooser is not permitted to incorporate random choices.

In a short while, the Computer will be given the choice between taking both boxes or Box #2 only. What will it do?

It turns out, we can easily predict the computer's decision. All we need to do is get a copy of the computer's program, run it on a different computer, and observe what the program does. This will give a nearly perfect prediction (unless something goes wrong when we load the program or there is some sort of glitch in the computer's operation). Then, when the actual game is played, we can witness our prediction come true. Virtually without exception, programs that pick both boxes are only awarded $1,000 while programs that only take Box #2 get the million.

In fact, most computer programs that appear to include randomness use a pseudorandom number generator. This is a bit of computer code that appears to generate random values but, in fact, follows a deterministic algorithm. See Chapter 21 to see how a deterministic process can give unpredictable results.

If we were asked to design a computer program to play this game, our decision would be utterly clear. Here's the program:

```
print("I take Box 2 only.")
```

When our program is run, our computer gets the million dollars and we're happy.

There's no reason to make the take-both-boxes decision. Because the action of a computer is knowable in advance (because it's purely mechanistic), this strategy will only net $1,000.

Why is there a contradiction when a human plays the role of Chooser, but not when the Chooser is a computer? Newcomb's paradox suggests that if we have free will, no Predictor can accurately forecast our actions.

Solution to puzzle on page 266: Let p be the probability that the Predictor is correct. The expected payouts are as follows:

$$\text{Box \#2 only} \to 1000000p$$
$$\text{Both boxes} \to 1000p + 1001000(1 - p).$$

We want the p that makes these equal to each other. That is, we solve the equation

$$1000000p = 1000p + 1001000(1 - p).$$

A bit of algebra gives the answer

$$p = \frac{1001}{2000} = 0.5005,$$

and so the cutoff is at 50.05%.

Further Reading

Petr Beckmann. *A History of π*. St. Martin's Press, third edition, 1976.

Edward B. Burger and Michael Starbird. *The Heart of Mathematics: An Invitation to Effective Thinking*. Wiley, third edition, 2009.

Underwood Dudley. *A Budget of Trisections*. Springer, 1987.

Richard P. Feynman, Robert B. Leighton, and Matthew Sands. *The Feynman Lectures on Physics*. Addison-Wesley Publishing Company, 1963.

Martin Gardner. Free will revisited, with a mind-bending prediction paradox by William Newcomb. *Scientific American*, 229, July 1973.

G. H. Hardy. *A Mathematician's Apology*. Cambridge University Press, 1940.

H. E. Huntley. *The Divine Proportion*. Dover, 1970.

Nicholas D. Kazarinoff. *Ruler and the Round: Classic Problems in Geometric Constructions*. Dover, third edition, 2003.

Mario Livio. *The Golden Ratio: The Story of PHI, the World's Most Astonishing Number*. Broadway, 2003.

Paul Lockhart. *A Mathematician's Lament: How School Cheats Us Out of Our Most Fascinating and Imaginative Art Form*. Bellevue Literary Press, 2009.

Eli Maor. *e: The Story of a Number*. Princeton University Press, 2009.

Paul J. Nahin. *An Imaginary Tale: The Story of $\sqrt{-1}$*. Princeton University Press, 2010.

James R. Newman. *The World of Mathematics*. Simon & Schuster, 1956. This four-volume collection is also available from Dover.

Mark Nigrini. *Benford's Law: Applications for Forensic Accounting, Auditing, and Fraud Detection*. Wiley, 2012.

E. Arthur Robinson and Daniel H. Ullman. *A Mathematical Look at Politics*. CRC Press, 2010.

Edward R. Scheinerman. *Mathematics: A Discrete Introduction*. Brooks/Cole, third edition, 2012.

M. L. Wantzel. Recherches sur les moyens de reconnaître si un problème de géométrie peut se résoudre avec la règle et le compas. *Journal de Mathématiques Pures et Appliquées*, 1(2):366–372, 1837.

Eric W. Weisstein. Mathworld—a Wolfram web resource. http://mathworld.wolfram.com/.

Index

A Mathematician's Apology, xvi, 1
absolute value, 158–162
accounting, forensic, 115
Adleman, Leonard, 17
Al-Khwarizmi, 124
aleph, 81
algebraically complete, 50
algorithm, 124–139
   Euclid, 136, 139
   recursive, 130
almost decisive, 249
ALPHABETICAL, 246
angle
   trisector, 151
angle trisection, 39
antinomy, Russell's, 83
antiprism, 188, 189
Archimedean solid, 188
area, 145–148
Arrow
   Impossibility Theorem, 256
   Kenneth, 256
axiom, 204
axiomatic set theory, 83

bank interest, 61–64
base
   sixteen, 26
   ten, 25, 27
   three, 26
   two, 21
Bayes' formula, 226–228
Benford
   Frank, 112
   law, 109–123
Bhaskara, 157
bijection, 72
binary, 19–26
Binet, Jacques, 100
bisection, 39
Bonacci, Leonardo, 86
BORDA, 253
Borda, Jean-Charles de, 254
box counting, 194–198
Brahe, Tycho, 1
bubble sort, 127

candidate neutrality, 247
Cantor, Georg, 71
cardinality, 72
carpet, Sierpinski, 200
center, 148–151, 165
   circumcenter, 150, 154
   incenter, 150
   of mass, 149
   orthocenter, 149
centroid, 148
chaos, 237
CIA World Fact Book, 110
circle
   definition, 165
   inscribed, 150
   kissing, 170
   packing, 168
   squaring, 39
circumcenter, 150, 154, 167
Cohen, Paul, 84
Collatz
   conjecture, 242–244
   Lothar, 244
complex number, 47–51, 158–162
composite, 8
congruent, 176
conjecture
   Collatz, 242–244
   Goldbach, 16
   Kepler, 169
   twin primes, 16
constructible number, 38
continuum hypothesis, 84
converge, 30
countable infinity, 81
cryptography, 17–18
cube, 176, 178, 179, 184, 187, 189
   doubling, 39
curvature, 170

decimal notation, 21, 27–32
decision method, 246
democracy, 245
Descartes, René, 170, 209
dice

Efron's, 220–222
  nontransitive, 217–222
  twenty-sided, 57
DICTATOR, 246, 250
dimension, 192–200
disk, 165, 195–197
  Poincaré, 210
divide and conquer, 127
divisible, 8
divisor, 8
  greatest common, 131–136
dodecahedron, 176, 178, 179, 184, 185, 187
domino, 86
doubling the cube, 39
duality, 178
dynamical system, 235

e, 60–70, 107
Efron, Bradley, 220
empty product, 10, 108
empty set, 82
equal temperament (intonation), 42
equilateral triangle, 151, 152
Escher, M. C., 213
Euclid, 11, 132, 136, 139, 204
Euclidean geometry, 208
Euler
  formula, 69, 175, 181
  Leonhard, 60, 178, 180
  number, see e
  polyhedral formula, 178–183
expected value, 264–265

factor, 8
factorial, 56, 60, 102–108
  zero, 107–108
factoring, 9–10
factorion, 107, 108
Fano plane, 208
Fermat
  Last Theorem, 162

Pierre de, 155, 162
Feynman, Richard, 7
Fibonacci, 86
  numbers, 88–101
forensic accounting, 115
fractals, 190–203
Fraenkel, Abraham, 83
free will, 266–268
function, 231–233
Fundamental Theorem
  of Algebra, 50–51
  of Arithmetic, 10, 36, 37

Garder, Martin, 259
Garfield, James, 157
Gaussian integer, 161
gcd, see greatest common divisor
geometry
  Euclidean, 208
  hyperbolic, 210
Gödel, Kurt, 84
Goldbach
  Christian, 16
  conjecture, 16
golden mean, 97
graph, 181
greatest common divisor, 131–136

Hales, Thomas, 169
Hardy, G. H., xvi, 1
hat check clerk, 64–66
Hero's formula, 146–147
hexadecimal, 26
Hexagon Theorem, 173
Hilbert, David, 84
hyperbolic geometry, 210
hypotenuse, 155

i, 44–51
icosahedron, 57, 176, 178–180, 184, 187
  stellated, 178
  truncated, 188

imaginary number, 46–47, 69
incenter, 150
independence of irrelevant alternatives, 255–258
induction, 91–93
infinity, 71–85
  countable, 81
integer, 8
  Gaussian, 161
interest, bank, 61–64
intonation
  equal, 42
  just, 41
  Pythagorean, 41
irrational number, 37, 54
isosceles triangle, 155
iteration, 231, 232

Jeopardy!, 94
just intonation, 41

Keats, John, 1
Kepler, Johannes, 169
Kepler, Johannes, 1
"Kiss Precise", 172
kissing circles, 170
Koch
  Helge von, 202
  snowflake, 202–203

law of logarithms, 123
lcm, see least common multiple
leaf, 182
least common multiple, 136–139
leg, 155
limit, 30
line, 207
  hyperbolic, 210
Lobachevsky, Nikolai, 207
logarithm, 120–123
  law of, 123
logarithmic scale, 68

logistic map, 233–237
lottery
   Pick Three, 264–265

MAJORITY, 246
mantissa, 115
mathematical induction, 91–93
May
   Kenneth, 249
   theorem, 249
merge sort, 127–131
MINORITY, 247
monotonicity, 248
Morley's Theorem, 152
most significant digit, 109
multiplication table, 112–115

naive set theory, 82
Newcomb
   paradox, 259–268
   Simon, 112
   William, 259
nontransitive dice, 217–222
notation
   scientific, 115–117
Nozick, Robert, 259
number
   complex, 47–51, 158–162
   constructible, 38
   Euler's, *see e*
   imaginary, 46–47, 69
   irrational, 37, 54
   rational, 33–34
   real, 44
   transcendental, 55, 60
   transfinite, 80–81
   triangular, 105–106
number theory, 17, 162

obtuse triangle, 154
octahedron, 176, 178, 179, 184, 186, 189
ODD, 247

Ode on a Grecian Urn, 1
oiler, *see* Euler
orthocenter, 149

π, 52–59, 107
packing circles, 168
paradox
   Newcomb, 259–268
Parallel Postulate, 206
Pascal
   Blaise, 173
   Hexagon Theorem, 173
   triangle, 201–202
Pi Day, 52
Pick Three lottery, 264–265
Pick's Theorem, 147–148, 152–154
place-value system, 20
plane
   Fano, 208
   hyperbolic, 210
Platonic solid, 175–187
PLURALITY, 253
Poincaré, Henri, 210
poker, 221–222
polygon, 175
polyhedron
   regular, 175–187
   semiregular, 187
polynomial, 50
Ponzi scheme, 129
postulate, 204
   Parallel, 206
preference profile, 245, 252
prime, 8–18
   gaps, 66–69
   number theorem, 66
   relatively, 56, 136
   twin, 16
prism, 175, 178, 179, 188, 189
probability
   conditional, 226–228
   dice, 217–222
   medical, 223–229

poker, 222
proof, 1–4
   by contradiction, xv, 13, 206
   by induction, 91–93
   combinatorial, 93–97
   geometric dissection, 156–158
pseudorandom number generator, 267
Ptolemy's Theorem, 167
public key cryptosystem, 17
pyramid, 175, 176, 178
Pythagoras, 39, 40, 155
Pythagorean
   scale (intonation), 41
   Theorem, 53, 59, 155–158, 166
   triple, 159–162

QED (*quod erat demonstrandum*), 4
quadratic formula, xvii

$\mathbb{R}$, 44, 72
radius, 165
radix point, 26
ranking, 217–222
rational number, 33–34
real number, 44
reciprocal, 48
rectangle, 212
recursive, 130
regular polygon, 175
regular polyhedron, 175–187
relatively prime, 56, 136
right triangle, 155, 167
Rivest, Ron, 17
Roman numerals, 19
round-off error, 239
RSA, 17
Russell
   antinomy, 83
   Bertrand, 83

Scarecrow, 155
scientific notation, 115–117
semiregular polyhedron, 187
sensitive dependence on initial conditions, 239
set, 71–72
  empty, 82
set theory
  axiomatic, 83
  naive, 82
Shamir, Adi, 17
Sierpinski
  carpet, 200, 203
  triangle, 190–192, 198–202
similar triangles, 212
snowflake, Koch, 202, 203
Soddy, Frederick, 172
sorting, 125–131
  bubble, 127
  merge, 127–131
sphere, 165
square root, 34
squaring the circle, 39
stellated icosahedron, 178
Stirling
  formula, 106–107
  James, 106
subset, 80

telescoping sum, 67
ternary, 26
tessellation, 212
tetrahedron, 176, 178, 179, 184, 186
Texas Hold 'Em, 221–222
theorem, 1–4, 10
  Arrow's impossibility, 256
  Fermat's Last, 162
  Fundamental
    of Algebra, 50–51
    of Arithmetic, 10, 36
  Hexagon, 173
  infinitely many primes, 11
  May, 249
  Morley, 152
  Pick, 147–148, 152, 154
  Prime Number, 66
  Ptolemy, 167
  Pythagorean, 53, 59, 155–158, 166
theory, 10
transcendental number, 55, 60
transfinite, 80–81
triangle, 143–154
  area, 145–148
  centers, 148–151

  centroid, 148
  equilateral, 151, 152
  isosceles, 155
  obtuse, 154
  right, 155, 167
  Sierpinski, 190–192, 198–202
  similarity, 212
triangular number, 105–106
trisection, 39
trisector, 151
truncated icosahedron, 188
twin primes conjecture, 16

unary, 20–21
uncountable, 81
unit, 8

VOTE-FOR-TWO, 253
voter neutrality, 246

Wantzel, Pierre, 40
Wiles, Andrew, 163
Wizard of Oz, 155
worst case analysis, 126

$\mathbb{Z}$, 8, 72
Zermelo, Ernst, 83
ZF Axioms, 83